SpringerBriefs in Physics

T0213855

For further volumes:
http://www.springer.com/series/8902

Ignazio Licata · Davide Fiscaletti

Quantum Potential: Physics, Geometry and Algebra

 Springer

Ignazio Licata
Davide Fiscaletti
Institute for Scientific Methodology
Palermo
Italy

ISSN 2191-5423   ISSN 2191-5431   (electronic)
ISBN 978-3-319-00332-0   ISBN 978-3-319-00333-7   (eBook)
DOI 10.1007/978-3-319-00333-7
Springer Cham Heidelberg New York Dordrecht London

Library of Congress Control Number: 2013953227

Printed on acid-free paper

Springer is part of Springer Science+Business Media (www.springer.com)

Foreword

It is extremely satisfying to see how the original work of de Broglie and Bohm is now being explored more fully from a number of different perspectives, leading to fresh insights into quantum phenomena and showing that the conventional interpretation is limited.

First came the claim that it was "impossible" to explain quantum phenomena in terms of individual particle movements. Then, when this was shown to be incorrect, the criticism changed to "this teaches us nothing new. Why add the metaphysical baggage of actual particles when the Bohm momentum $pB = \nabla S$ and the kinetic energy cannot be measured." But even this criticism has now been shown to be wrong with the appearance of weak measurements.

Rick Leavens and others have shown that the weak value of the momentum operator at a post-selected position *is* the Bohm momentum and that the weak value of the operator $\widehat{P}^2(x,t)/2m$ *is* the Bohm kinetic energy. The quantum potential is now open to experimental investigation. Numbers can be attached to these quantities which can then be compared with experiment. Recently, experiments of this kind have been carried out on photons by Aephraim Steinberg and his group in Toronto. More experiments are in progress, this time to make weak measurements on atoms, bringing the experiments even closer to verifying the predictions of the extra phenomena revealed through the de Broglie–Bohm approach.

Now we appear to have a more believable explanation of quantum phenomena, but it still leaves us with one feature that many find objectionable, namely, quantum nonlocality. This opposition is maintained in spite of all the theoretical and experimental support for some form of nonlocality. It leaves us with the question of how we are to understand the resulting tension that exists between local relativity and quantum nonlocality.

Experiments show that the nonlocal correlations turn out to be correct even when detection events are space-like separated. Of course we should not be surprised at these results because Bohr had already argued that there was a novel kind of wholeness involved in all quantum phenomenon. He talks of the "impossibility of making any sharp separation between the behaviour of atomic objects and the interaction with measuring instruments which serve to define the conditions under which the phenomena appear." When replying to EPR, he talks about "an influence on the very conditions which define the possible types of predictions regarding the future behaviour of the system." An *influence* acting over space-like separations? What *influence*? Could it be the quantum potential?

Although the de Broglie–Bohm approach allows a separation between particles, it is the quantum potential that "locks" them together. Furthermore, this coupling does not appear to be propagated as in a classical interaction. Rather it appears as a global constraint on the whole process that begins to make Bohr's claims of "wholeness" clearer. Could these global constraints, which are reflected in the covering groups of the symmetries involved, be the real explanation of these apparently nonlocal effects? More investigations are needed and this is where this book can play a valuable role. Ignazio Licata and Davide Fiscaletti have collected together a number of substantial explorations of the de Broglie–Bohm model, particularly those that explore possible meanings of the quantum potential. This gives us a valuable source of relevant material which can play an important role in taking these investigations further into new areas.

17 October 2013   B. J. Hiley

Preface

Who Needs Quantum Mechanics Interpretations?

The activity on Quantum Mechanics "interpretations" is still a flourishing industry cranking out new products every change of season. Like any production activity, it shows many self-referential aspects tending to wear out criticality the logistic curve of economists. Most of such research is justified by the longstanding—and unsolvable—problems of Copenhagen interpretation, such as the physical nature of the wave function, the collapse problem, and the observer role.

And yet, if we look outside of what H. Pagels called "the bazaar of reality," that is the actual use of Quantum Physics in real physics problems, we will see that something significant has happened for "foundations" in these years.

For example, the 2012 Nobel prize has been awarded to David Wineland and Serge Haroche, whose by far classical experiments in quantum optics has showed the full citizenship of those "strange"—and so embarrassing for the Standard Interpretation—negative probabilities of Wigner–Feynman distribution. It is just of negative probabilities that Feynman, with his usual intellectual audacity, has talked about in his contribution in honor of David Bohm [1].

The development of quantum cosmology has incisively proposed again the Robert Serber statement: "Quantum Mechanics laws are applied to Big-Bang now, but there were no observers then!" [2], so driving the metaphysics of the observer more and more toward Alice's wonderland. After all, restarting from Pascual Jordan and Werner Heisenberg, Marcello Cini, one of the masters of Italian theoretical physics, wrote: "I know very well we can't come back and physicists will keep on looking at Schrödinger equation as a fundamental pillar of their subject, as for me, after having teach it for 50 years, I'm happy I realized that wave functions can go and take their place next ether at the junk store" [3]. He completely developed such magnificent program in a 2003 paper; here we quote a key statement from the abstract: "In this formulation the wave/particle duality is no longer a puzzling phenomenon. The wave/particle duality is instead, in this new perspective, only the manifestation of two complementary aspects (continuity versus discontinuity) of an intrinsically nonlocal physical entity (the field) which objectively exists in ordinary three-dimensional space" [4].

On an opposite and a complementary front, G. Preparata stated: "The unacceptable subjectivism permeating the generally accepted Quantum Mechanics (QM) interpretation based on Niels Bohr ideas and Copenhagen school makes sense only in understanding it is not a complete theory of reality (...) To complete it we have to put it aside and admit Quantum Field Theory (QFT) as the only description of reality. Quantum Mechanics is so just an approximation of QFT and it is limited to the analysis of quantum processes in space-time regions where, in all probability, we can find a single quantum with its respective wave field" [5, 6]. The recent Afshar experiment [7] seems to confirm what Cini and Preparata said; it does not "demolishes" nor does it "elude" Heisenberg Uncertainty Principle, but places it within its natural version in QFT, which links the number of quanta to phase. In short, as J. Cramer says, a real "farewell to Copenhagen" [8], which gave birth to strong processes of critical revision starting just from that 1927 Solvay Conference where it become the standard interpretation [9, 10]. Obviously, an amazing series of experiments has been crucial in showing that nonlocality is neither something mysterious buried under the statistical machinery of probabilistic interpretation nor the Einstein spooky action at a distance, but a central element of the physical world' logics.

Such processes should not be possible without an increasing acceptance of the David Bohm Quantum Physics vision. Today, it can be considered, at least For All Practical Purpose! (FAPP) as the dominant reading of Quantum Physics, able to reopen many questions which were considered closed or forever "paradoxical". For the most of physicists, thinking in Bohmian terms is simpler than thinking within the restricted vetoes of the Copenhagen vulgate. We recall that it was just D. Bohm who changed the "ideal" EPR experiment in an operative procedure which experimental physicists uses largely still today and that John Bell, by starting from Bohm work, developed his famous inequality, so becoming one of the most brilliant supporter of Bohmian "speakability" of QM [11]. The constant and distinctive element of Bohm's work has been the crucial role of nonlocality to be introduced ab initio in the structural corpus of the physical theory, and quantum potential, even in the plurality of the mathematical treatments, is in the center of such structure, the nonlocal trait d'union between the post-classical features of Quantum Physics and the most advanced perspectives of Field Theory [12, 13].

For a long time, Bohm's one has been "just" another interpretation, and almost till the day before yesterday it was tout court identified with the de Broglie-Bohm Pilot Wave Theory. Actually, the idea of a particle surfing a wave is typical of de Broglie who felt the need to recover somewhat the classical concept of trajectory, still today a sort of dogma motion central in the work of exceptional and very active Bohmians [14, 15]. The fact that Bohm's ideas were considerably different from classical and post-classical ones is witnessed by this excerpt from the "duel" with M. Pryce, a defender of the orthodox vision, broadcasted by BBC in 1962:

"We wondered what actually an electron does. What should it do while it passes from the source to the slit? That's the point. Well, I could propose, for example, that the electron is not a particle in the sense it is currently meant, but an event. I assume such event happens in a generic medium—a "field"—we can suppose in this field there's an impulse. A wave moves forward and converges in a point so producing a very strong impulse and then diverges and scatters away. Let's imagine these impulses in a series all reaching a line there producing a series of intense pulses. The impulses will be very close one to the other, and so they will look like a particle. In most cases, all that will behave just like a particle and will behave differently when goes through the two slits, because each impulse will come out according to the way the incident wave passes the slits. The result is that we are looking at something it's neither a wave nor a particle. If you wonder how the electron has actually passed the slit and if it has really passed one slit or the other, I would reply that probably is not that kind of thing which can pass a slit or the other one. Actually, it is something which forms and dissolves continuously and that can be the way it really acts" [16].

His collaboration with J. P. Vigier, the most famous and brilliant among de Broglie disciples, will last for a long time (very strong between 1954 and 1958, it become progressively less intense after 1963); the French scientist will pursue his idea of a quantum stochastic geometrodynamics where particles are like solitons in nonlinear fields and nonlocality is a form of superluminality [17]. Bohm, convinced that this approach is not sufficiently radical, will instead follow the algebraic-topological line which characterizes his intense last years work with Basil Hiley. The revolutionary idea is that nonlocality does not look like a field at all, but it is "written" in the informational structure of a pre-space that Bohm-Hiley called "Implicate Order". This one is revealed only partially, depending on the information the observer chooses to extract from the system, and gives to QM its characteristic "contextuality" [18]. It is the recovery of the old Bohr complementarity, now based not on something "elusive" and the "uncertain" role of the observer, but on the deep logic of the physical world and the noncommutative relation between system and environment. It is not only a matter of re-reading the wave function as a statistical covering of a great number of transitions between the field modes. The theory of Implicate/Explicate order is the first real attempt to realize the J. A. Wheeler program of It from Bit (or QBit), the possibility to describe the emergent features of space-time-matter as expressions, constrained and conveyed, of an informational matrix "at the bottom of the world" [19].

Beyond any more or less ephemeral trends, this is the research of the great climbers of theoretical physics, such as Basil Hiley and David Finkelstein.

Anyway, the most prevailing reading of the Bohm work is still that known as the de Broglie-Bohm Wave Pilot Theory, surely for its intuitive advantages and the relatively simple formalism. We could say that Hiley and his collaborators are Bohmist rather that Bohmian, in the same sense as the word Marxian is opposed to Marxist. Everybody agree on the fact that beyond Copenhagen Quantum Theory can produce good Physics without (bad) Philosophy. Even the longstanding question of the "realism" of trajectories seems to have found its turning point in the connections with the Feynman Paths and the pregeometries [for a review, see: 20–22]. Thus, the quantum potential showed to be the most powerful, inclusive, and flexible tool "to tell" the nonlocal quantum processes, so opening new perspectives in Field Theory, Particle Physics, Gravitation Theory, Cosmology,

Quantum Information and Chaos, and laying the foundations for a grounded bridge able to unify QM and QFT. Obviously, there are many ways "to read" the quantum potential, and it is worthy to be underlined that the differences more that interpretative are linked to the problem under consideration. There are semiclassical approaches, stochastic ones, geometrodynamic ones up to the purely algebraic and topological ones. The subtitle of this short book (Physics, Geometry, and Algebra) suggests we tried to provide the reader with complete review of the different positions into play: from the wave pilot, into subquantum thermodynamics and stochastics up to the noncommutative geometries and Clifford Algebras. The heedful reader will maybe able to reveal some little, subtle dissonances here and there, due to the fecund and sometimes fierce debating between the authors.

The prefacer (IL) is definitely a Bohmist, whereas my co-author is more at ease with quantum geometrodynamics. What come out in the end is not what Piotr Garbaczewski defines as quantum political/quasi-religious parties but—at least we hope so—a careful and punctual survey of the main results of a growing bibliography, updated till the very moment I write such lines.

We have neither authorial aspirations nor the pretense to replace the D. Bohm, B. Hiley great classical books, or the P. Holland and D. Dürr, S. Goldstein, e N. Zanghì treatises. As for the applications we address the reader to the monumental Applied Bohmian Mechanics [23].

We hope to give a light introductive guide to what can be actually done by the Bohmian and Bohmist tools and to contribute to the awakening from that "dogmatic sleep" about the mysteries of the wave function, a direction that seems to be followed also by the recent operational trend of quantum Bayesism [24].

The quantum potential not only explains the probabilistic nature of the wave function, but is an open door into the deep informational structure of the Universe.

Ignazio Licata

References

1. Hiley, B., Peat, F.D. (eds.): Quantum implication. Essays in honour of David Bohm. Psychology Press, Routledge (1987)
2. Serber, R., Crease, R.P., Mann, C.C.: The second creation: Makers of the revolution in twentieth-century physics. Rutgers University Press, New Brunswick (1996)
3. Cini, M.: Personal communication, Feb (2002)
4. Cini, M.: Field quantization and wave particle dualism. Ann. Phy. **305**, 83–95 (2003)
5. Preparata, G.: Dai quark ai cristalli. Bollati Boringhieri, Torino (2002) (in Italian)
6. Preparata, G.: An introduction to a realistic quantum physics. World Scientific, Hackensack (2002)
7. Afshar, S.S.: Violation of Bohr's complementarity: One slit or both? AIP Conference Proceedings **810**, 294–299 (2006)
8. Cramer, J.: A farewell to Copenhagen? Analog, Dec (2005)

9. Bacciagaluppi, G., Vantini, A.: Quantum theory at the Crossroads: Reconsidering the 1927 Solvay Conference. Cambridge University Press, Cambridge (2009)
10. Valentini, A.: Beyond the quantum. Phy. World, 32–37 (2009)
11. Bell, J.: Speakable and unspeakable in quantum mechanics, 2nd edn. Cambridge University Press, Cambridge; (2004)
12. Licata, I.: Osservando la sfinge, Di Renzo Editore, Rome (2009) (in Italian)
13. Fiscaletti, D.: I gatti di Schrödinger, Muzzio Editore, Rome (2007) (in Italian)
14. Holland, P.: The quantum theory of motion: An account of the De Broglie-Bohm causal interpretation of quantum mechanics. Cambridge University Press, Cambridge (1993)
15. Dürr, D., Goldstein, S., Zanghì, N.: Quantum physics without quantum philosophy. Springer, New York (2013). Bohmian-Mechanics.net
16. Toulmin S. (ed.): Reported in quanta and reality. Hutchinson, London (1971)
17. Hunter, G., Jeffers S., Vigier J.P. (eds.): Causality and locality in modern physics: Proceedings of a symposium in honour of Jean-Pierre Vigier. Springer, New York; Softcover reprint of hardcover 1st ed. 1998 edition (2010)
18. Licata, I.: Emergence and computation at the edge of classical and quantum systems. In: Licata. I., Sakaji A. (eds.) Physics of Emergence and Organization. World Scientific, Singapore (2008)
19. Wheeler, John A.: Information, physics, quantum: The search for links: In Zurek W. (ed.) Complexity, Entropy, and the Physics of Information, Redwood City. Addison-Wesley, California (1990)
20. Boscá, M.C.: Some observations upon "realistic" trajectories in Bohmian quantum mechanics. Theoria **76**, 45–60 (2013)
21. Hiley, B.J., Callaghan, R.E., Maroney, O.: Quantum trajectories, real, surreal or an approximation to a deeper process? arXiv:quant-ph/0010020v2
22. Licata, I., Fiscaletti, D.: Bohm trajectories and Feynman paths in light of quantum entropy. Centr. Europ. J. Phys. (in press)
23. Oriols, X., Mompart, J., Cirac, I.: Applied Bohmian mechanics: From nanoscale systems to cosmology. Pan Stanford Publishing, Singapore (2012)
24. von Baeyer, H.C.: Quantum weirdness? It's all in your mind. Sci. Amer. June (2013)

Contents

Chapter 1
A Short Survey on a "Strange" Potential

1.1 From de Broglie's Pilot Wave to Bohm's Quantum Potential

The David Bohm work on quantum mechanics starts from de Broglie pilot wave theory, which can be considered as the most significant hidden variables theory equivalent to quantum mechanics as for predictability and able to restore a causal completion to quantum mechanics. Such approach was originally proposed by Louis de Broglie at Solvay Conference in 1926. As regards the non-relativistic problem, de Broglie suggested that the wave function of each one-body physical system was associated with a set of identical particles which have different positions and are distributed in space according to the usual quantum formula, given by $|\psi(\vec{x})|^2$. But he recognized a dual role for the wave function: on one hand, it determines the probable position of the particle (just like in the standard interpretation); on the other hand, it influences the position by exerting a force on the orbit. According to de Broglie, the wave function would act like a sort of pilot wave which guides the particles in regions where such wave function is more intense [1–3]. In the context of his proposal, de Broglie applied his guidance formula to calculate the orbits for the hydrogen atom stationary states.

De Broglie gave a particular beautiful explanation of how the pilot wave would actually guide the particle. He proposed that inside the particle there was a periodic process equivalent to a clock. In a frame at rest, the clock would have a frequency $\omega_0 = m_0 c^2 / \hbar$ where m_0 is the rest mass of the particle. By considering how this clock would behave in other frames, he derived what is in essence the Bohm-Sommerfeld relationship $\oint \vec{p} \cdot dx = nh$ which is the condition for the clock to remain in phase with the pilot wave. The nonlinear interaction describing the phase locking of synchronous motors can be compared to a similar nonlinear interaction regarding de Broglie's further development of the model of pilot wave known as "double solution theory". What de Broglie also suggested was that nonlinearity made possible a singular solution representing the particle, which had to fit smoothly onto a weak background field, obeying Schrödinger's equation as a

I. Licata and D. Fiscaletti, *Quantum potential: Physics,*
Geometry and Algebra, SpringerBriefs in Physics,
DOI: 10.1007/978-3-319-00333-7_1, © The Author(s) 2014

linear approximation. This requirement of a smooth connection of the two members of the double solution was what explained the guidance condition.

De Broglie approach met with a general lack of enthusiasm. In particular, Pauli [4] made important objections, based on the fact that de Broglie's approach did not provide a consistent account for many-body systems (for example, a two-body scattering process). While de Broglie felt that he had at least the germ of an answer, this was apparently not appreciated by the majority of the physicists there present. This, along with the fact that not even Einstein spoke up for the theory, led to a definite rejection of the theory. The unfavourable climate, presumably determined by Heisenberg's discovery of the 'uncertainty' relations, eventually led de Broglie to abandon his programme. De Broglie returned to research in this field 25 years later when David Bohm rediscovered the approach and developed it to the level of a fully fledged physical theory.

In his two papers of 1952 *A suggested interpretation of the quantum theory in terms of hidden variables* [5, 6]—which can be considered the starting point of the de Broglie-Bohm theory—Bohm succeeded in extending coherently the de Broglie's approach. In particular, in the first of these papers shows to provide a generally consistent account of quantum mechanics for the one-body systems. In the second paper, this work was extended to the many-body systems, in a way that answered precisely and coherently Pauli's objections, and made possible key new insights into the meaning of quantum mechanics.

Bohm's version of quantum mechanics is practically the de Broglie pilot-wave theory carried to its logical conclusion. The essential insight of Bohm's work is the interpretation of quantum mechanics as a theory about particles in motion: in addition to the wave function, the description of a quantum system also includes its configuration—that is to say, the precise positions of all the system's particles at all times. It is just for this reason that Bohm's theory is often called a hidden variables theory: it is based on the idea that quantum mechanics is not complete and must be completed by adding supplementary parameters to the formalism. The hidden variables of the model are just the positions of all the particles constituting the physical system under examination: the physical system is prepared in such a way that, at the initial time $t = 0$, it is associated with a specific wave function $\psi(\vec{x}, 0)$ which is assumed to be known perfectly, and moreover is in a point $\vec{x}$ (among those compatible with the wave function) that instead we ignore.

We can express the core of 1952 Bohm work by the following basic postulates:

1. An individual physical system comprises a wave propagating in space and time together with a point particle which moves continuously under the guidance of the wave.
2. The wave is mathematically described by $\psi(\vec{x}, t)$, a solution to Schrödinger's equation

$$i\hbar \frac{\partial \psi}{\partial t} = H\psi \qquad (1.1)$$

(where H is the Hamiltonian of the physical system and $\hbar$ is Planck's reduced constant).

3. The particle motion is obtained as the solution $\vec{x}(t)$ to the equation

$$\dot{\vec{x}} = \frac{1}{m} \nabla S(\vec{x}, t)\big|_{\vec{x}=\vec{x}(t)} \tag{1.2}$$

where S is the phase of $\psi(\vec{x}, t)$ and m is the mass of the particle. This equation can be solved by specifying the initial condition $\vec{x}(0) = \vec{x}_0$. This specification constitutes the only extra information introduced by the theory that is not contained in $\psi(\vec{x}, t)$ (the initial velocity is fixed once one knows S). An ensemble of possible motions associated with the same wave is generated by varying $\vec{x}_0$.

These three postulates on their own constitute a consistent theory of motion. In order to guarantee the compatibility of the motion of the ensemble of particles with the results of quantum mechanics, Bohm's introduces a further postulate:

4. The probability that a particle in the ensemble lies between the points $\vec{x}$ and $\vec{x} + d\vec{x}$ at time t is given by $R^2(\vec{x}, t)d^3x$ where $R^2 = |\psi|^2$ (namely R is the real amplitude function of $\psi(\vec{x}, t)$).

This fourth postulate has the effect to select among all the possible motions implied by the law (1.2) those that are compatible with an initial distribution $R^2(\vec{x}, 0) = R_0^2(\vec{x})$. As regards the postulate 4, it is also important to remark that $|\psi(\vec{x}, t)|^2$ determines the probability density of finding a particle in the volume d^3x at time t if a suitable measurement is carried out.

In Bohm's approach to non-relativistic quantum mechanics, the introduction of the particle motion occurs by substituting the expression of the wave function in polar form $\psi = R\, e^{iS/\hbar}$ into Schrödinger's equation. By separating the real and the imaginary part of Schrödinger's equation, Bohm showed that, for one-body systems, the movement of the corpuscle under the guide of the wave happens in agreement with a law of motion which assumes the following form

$$\frac{\partial S}{\partial t} + \frac{|\nabla S|^2}{2m} - \frac{\hbar^2}{2m}\frac{\nabla^2 R}{R} + V = 0 \tag{1.3}$$

(where V is the classical potential). This equation is equal to the classical equation of Hamilton-Jacobi except for the appearance of the additional term

$$Q = -\frac{\hbar^2}{2m}\frac{\nabla^2 R}{R} \tag{1.4}$$

having the dimension of an energy and containing Planck constant and therefore appropriately defined quantum potential. The motion equation of the particle can be expressed also in the form

$$m\frac{d^2\vec{x}}{dt^2} = -\nabla(V + Q) \tag{1.5}$$

where $\vec{x} = \vec{x}(t)$ is the trajectory of the particle associated with its wave function. Equation (1.5) is equal to Newton's second law of classical mechanics, always with the additional term Q of quantum potential. The movement of an elementary particle, according to Bohm's pilot wave theory, is thus tied to a total force which is given by the sum of two terms: a classical force (derived from a classical potential) and a quantum force (derived just from the quantum potential) [7, 8].

Therefore, one can say that, according to de Broglie-Bohm 1952 theory, each subatomic particle is completely described by its wave function (which evolves according to the usual Schrödinger equation (1.1) and its configuration and follows a precise trajectory $\vec{x} = \vec{x}(t)$ in space-time that is originated by the action of a classic potential and a quantum potential (and that evolves according to the Eq. (1.3) or the equivalent Eq. (1.5)). Moreover, it is important to underline that in Bohm's approach to non relativistic quantum mechanics, the decomposition of the Schrödinger equation (1.1) leads, besides to the quantum Hamilton-Jacobi equation (1.3), also to a continuity equation for the probability density $\rho(\vec{x}, t) = R^2(\vec{x}, t) = |\psi(\vec{x}, t)|^2$:

$$-\frac{\partial \rho}{\partial t} = \nabla \cdot \left(\rho \frac{\nabla S}{m} \right). \tag{1.6}$$

Equation (1.6) show that all individual trajectories demonstrate collective behaviour like a superconductive flux.

In synthesis the fundamental equation of motion is the quantum Hamilton-Jacobi equation (1.3), with Schrödinger's equation (1.1), while the guidance equation ruling the evolution of the trajectories (1.2) may be regarded as a constraint to the more basic Eq. (1.3). The actual trajectory pursued by the physical system under consideration derives from the combined action of the classical potential and, above all, of the quantum potential (which measures its deviation from the classical trajectory) and emerges from the fundamental equation of motion (1.3).

The quantum potential must not be considered an ad hoc term inside quantum theory. In the formal structure of Bohm's non-relativistic theory, it emerges directly from Schrödinger's equation and without it energy should not be conserved. In fact, taking into account that the quantity $-\frac{\partial S}{\partial t}$ is the total energy of the particle and that $\frac{|\nabla S|^2}{2m}$ is its kinetic energy, equation (1.3) can be also written in the equivalent form

$$\frac{|\nabla S|^2}{2m} - \frac{\hbar^2}{2m} \frac{\nabla^2 R}{R} + V = -\frac{\partial S}{\partial t} \tag{1.7}$$

which can be seen as a real energy conservation law in quantum mechanics: here one can easily see that without the quantum potential (1.4) energy could not be conserved and this means that quantum potential plays an essential role in quantum formalism.

The treatment provided by relations (1.3), (1.4), (1.5), (1.6) and (1.7) can be extended in a simple way to many-body systems. If we consider a wave function $\psi = R(\vec{x}_1, \ldots, \vec{x}_N, t)e^{iS(\vec{x}_1, \ldots, \vec{x}_N, t)/\hbar}$, defined on the configuration space R^{3N} of a system of N particles, the movement of this system under the action of the wave ψ takes place in agreement to the law of motion

$$\frac{\partial S}{\partial t} + \sum_{i=1}^{N} \frac{|\nabla_i S|^2}{2m_i} + Q + V = 0 \tag{1.8}$$

where

$$Q = \sum_{i=1}^{N} -\frac{\hbar^2}{2m_i} \frac{\nabla_i^2 R}{R} \tag{1.9}$$

is the many-body quantum potential. Equation (1.8) can also be written in the form

$$\sum_{i=1}^{N} \frac{|\nabla_i S|^2}{2m_i} + Q + V = -\frac{\partial S}{\partial t} \tag{1.10}$$

which is the energy conservation law for a many-body system. The equation of motion of the i-th in particle, in the limit of big separations, can also be written in the following form

$$m_i \frac{\partial^2 \vec{x}_i}{\partial t^2} = -[\nabla_i Q(\vec{x}_1, \vec{x}_2, \ldots, \vec{x}_n) + \nabla_i V_i(\vec{x}_i)] \tag{1.11}$$

which is a quantum Newton law for a many-body system. Equation (1.11) shows that the contribution to the total force acting on the i-th particle coming from the quantum potential, i.e. $\nabla_i Q$, is a function of the positions of all the other particles and thus does not decrease with distance. The continuity equation for the probability density becomes:

$$-\frac{\partial \rho}{\partial t} = \sum_{i=1}^{N} \nabla_i \cdot \left(\rho \frac{\nabla_i S}{m} \right). \tag{1.12}$$

The mathematical expression of quantum potential shows that this entity has not the usual properties expected from a classic potential. On the basis of its definition (1.4) (or the analogous (1.9) for many-body systems), the quantum potential depends on how the amplitude of the wave function varies in space. The presence of Laplace operator indicates that the action of this potential is like-space, namely creates a non-local, instantaneous action onto the particle. In Eqs. (1.4) and (1.9), the appearance of the amplitude of the wave function in the denominator also explains why the quantum potential can produce strong long-range effects and the typical properties of entangled wave functions. This is just the kind of behaviour required to explain the EPR paradox.

If we examine the expression of the quantum potential in the two-slit experiment, we find that it depends on the width of the slits, their distance apart and the momentum of the particle. In other words, it has a contextual nature, namely brings a global information on the process and its environment. It contains an instantaneous information about the overall experimental set-up. This information modifies the global behaviour of the physical system. In a double-slit experiment if one of the two slits is closed the quantum potential changes, and this information arrives instantaneously to all the particles, which changes their final pattern. In this regard, in 1984 D. Bohm and Basil Hiley suggested that the quantum potential can be interpreted as a sort of "information potential": the particles in their movement are guided by the quantum potential just as a ship at automatic pilot can be handled by radar waves of much less energy than that of the ship. On the basis of this interpretation, the results of double-slit experiment are explained by saying that the quantum potential contains an "active information".

Moreover, as regards the properties of the quantum potential that determine a radical departure from classical Newtonian physics, Bohm and Hiley underlined that the quantum potential is not changed when the wave function is multiplied by an arbitrary constant and they commented on this as follows: "This means that the effect of the quantum potential is independent of the strength (i.e., the intensity) of the quantum field but depends only on its form. By contrast, classical waves, which act mechanically (i.e., to transfer energy and momentum, for example, to push a floating object), always produce effects that are more or less proportional to the strength of the wave" [9].

As regards the crucial role played by the quantum potential in the explanation and understanding of quantum phenomena, one can mention that in 1979 Philippidis, Dewdney and Hiley [10] studied in detail the double-slit experiment of the electrons showing that by means of Bohm's quantum potential it is possible to explain the experimental results (namely the interference pattern) maintaining the concept of trajectory and particle even after the electrons have passed the slits. The quantum potential can explain also the other typical quantum phenomenon represented by tunnelling effect through a square barrier, justifying in a clear way why some particles manage to pass the potential barrier and others not. On the basis of the calculations performed by Dewdney and Hiley in 1982, the tunnelling effect occurs as a consequence of a severe modification of the classical potential caused by the quantum potential: the quantum potential is negative in the region of the barrier and thus it can be possible for a particle to have sufficient energy to pass the barrier [11]. These results regarding double-slit interference and tunnelling—as well as the fact that during the 1980s Bohm's quantum potential approach provided a convincing explanation of many other experiments (in this regard the reader may find details, for example, in the Ref. [12])—show the legitimacy and physical coherence of the de Broglie-Bohm theory.

As regards the interpretation of non-relativistic quantum mechanics proposed by Bohm, a great deal of confusion exists about what Bohm was saying exactly when he first published his 1952 papers. Many discussions about Bohm's ideas in these two papers assume that they represented some kind of definitive theory.

As shown clearly by Hiley in the recent article *Some remarks on the evolution of Bohm' proposals for an alternative to standard quantum mechanics* [13], the misunderstandings about Bohm's original ideas, grew after the appearance of the term "bohmian mechanics" in a 1992 paper of Dürr, Goldstein and Zanghì [14]. In his classic works of 1952, Bohm never used the term mechanics: in these two papers Bohm's intention was not to find a classical order based on a deterministic mechanics from which the quantum formalism would emerge. Indeed the content of his book *Quantum Theory* published in 1951 [15], which gives an exhaustive account of the orthodox view of the theory, already shows the seeds of how radical a change Bohm thinks is needed in order to begin to understand the structure that underlies the quantum formalism. By reading that book one can see easily in Bohm's ideas the need to go beyond mechanical ideas. In the section of this book titled 'The need for a non-mechanical description', Bohm writes ....*the entire universe must, in a very accurate level, be regarded as a single indivisible unit in which separate parts appear as idealisations permissible only on a classical level of accuracy of the description. This means that the view of the world as being analogous to a huge machine, the predominant view from the sixteenth to nineteenth century, is now shown to be only approximately correct. The underlying structure of matter, however, is not mechanical* [15].

On the basis of Bohm's 1952 papers, the mathematical formalism synthesised in Eqs. 1.1–1.12 does not present an opposition to the quantum standard formalism: Eqs. 1.1–1.12 merely shows that an alternative view, that attributes definite properties to individual particles, is possible without radically changing the formalism and altering the predictions. He was not offering these proposals as the final definitive interpretation of the quantum formalism in the non-relativistic domain. On the other hand, throughout the papers he stresses that his approach opens up possibilities of modifying the formalism in ways that could not be possible in the standard interpretation. For example, he suggested that there was a possibility of exploring deeper structures that could lie below 10^{-13} cm. Clearly he was not presenting the ideas as being, in some sense, a definitive interpretation of the formalism, but using it to look for something deeper.

As shown clearly also by Hiley [13], the fundamental key element which shows that Bohm's 1952 approach was not mechanical is the quantum potential. In his 1952 papers, Bohm showed that, in virtue of the mathematical features of the quantum potential, the whole environment determines the properties of the individual particles and their relationship. In Bohm's view, the quantum potential implies a universal interconnection of things that could no longer be questioned. In Bohm's ideas, the notion of the particle cannot be considered more fundamental than the quantum potential. At any stage of the analysis there are quasi-stable, semi-autonomous features that could be related, ordered and melted into a coherent structure which we could call a theory. However these quasi-stable features, namely the 'particles', of the theory took their properties from the total process itself. What one can conclude from the analysis of Bohm's 1952 papers is that the quantum potential enables the global properties of quantum phenomena to be focused on the particle aspect, where the 'particle' is not independent of the background. Furthermore it is the quantum potential that contains the effect of this background. This implies that the particle and quantum potential form an indivisible whole.

1.2 The Dynamic and Geometric Approaches to the Quantum Potential

Since the 1990s of the last century we assist at a growing interest in Bohm's quantum potential approach. In particular, according to the modern research, there are two fundamental interpretations and keys of reading of the quantum potential: on one hand, the dynamic and geometric approaches and, on the other hand, the algebraic approaches. In this chapter we focus our attention on the dynamic and geometric approaches while the Sect. 1.3 will be dedicated to the algebraic approaches.

1.2.1 The Quantum Trajectories of Dürr, Goldstein, Tumulka and Zanghì

The first significant geometrodynamic approach to Bohm's quantum potential can be considered the dynamic approach developed by D. Dürr, S. Goldstein, R. Tumulka and N. Zanghì. Since 1992, when the article *Quantum equilibrium and the origin of absolute uncertainty* appeared in *The Journal of Statistical Physics*, Dürr, Goldstein and Zanghì suggested a re-reading of Bohm's interpretation based on the idea that well-defined Bohmian trajectories can be considered the foundation, and thus the primary physical reality of quantum mechanics. This re-reading of de Broglie-Bohm theory, which has been called as Bohmian mechanics by its authors, has been developed in the last 15 years thanks also to M. Esfeld, V. Allori and other authors. In order to analyse the most important features of Dürr's, Goldstein's, Tumulka's and Zanghì's approach, we refer to the papers [14, 16–25].

According to the dynamic approach of Dürr, Goldstein, Tumulka and Zanghì, in non-relativistic quantum mechanics a family of trajectories in configuration space is associated with the wave function under consideration and satisfies a guiding equation. The view of Dürr, Goldstein, Tumulka and Zanghì, claims that, in our world, the primary physical reality of electrons and other elementary particles is their trajectory which is governed by a guiding equation intended as the fundamental equation of motion. The guiding equation states that the fundamental object which influences the motion of the particle is the wave function. In other words, in the picture provided by Dürr, Goldstein, Tumulka and Zanghì, the fundamental equation of motion (1.3) of Bohm's original approach is replaced by a new equation of motion (namely the guiding equation) which does not talk about forces, about pushes or causes, but merely defines the trajectory in terms of the wave function. The quantum potential enters the description as a secondary element which serves, together with the guiding equation, to find the wave function determining the real, actual trajectories of the particles.

Therefore, let us consider a wave function $\psi(q)$ of non-relativistic quantum mechanics, defined on the configuration space R^{3N} of a system of N particles and which evolves according to the Schrödinger equation

$$i\hbar \frac{\partial \psi}{\partial t} = -\sum_{k=1}^{N} \frac{\hbar^2}{2m_k} \nabla_k^2 \psi + V\psi \tag{1.13}$$

where m_k is the mass of the k-th particle, V is the classical potential. In the Bohmian mechanics proposed by Dürr, Goldstein, Tumulka and Zanghì, this wave function $\psi(q)$ exerts as influence on the motion of the system by determining a family of trajectories in the configuration space R^{3N}, the Bohmian trajectories, which are defined as those trajectories $t \rightarrow Q(t) = \left(\vec{Q}_1(t), \ldots, \vec{Q}_N(t)\right)$ satisfying the guiding equation

$$\frac{d\vec{Q}_k(t)}{dt} = \frac{\hbar}{m_k} \text{Im} \frac{\nabla_k \psi}{\psi} (Q(t)). \tag{1.14}$$

In equivalent terms, the wave function $\psi(q)$ determines a vector field

$$v^\psi = \left(\vec{v}_1^\psi, \ldots, \vec{v}_N^\psi\right), \ \vec{v}_k^\psi = \frac{\hbar}{m_k} \text{Im} \frac{\nabla_k \psi}{\psi} \tag{1.15}$$

and Eq. (1.14) becomes

$$\frac{dQ(t)}{dt} = v^\psi(Q(t)). \tag{1.16}$$

By defining the probability current

$$j^\psi = \left(\vec{j}_1^\psi, \ldots, \vec{j}_N^\psi\right), \ \vec{j}_k^\psi = \frac{\hbar}{m_k} \text{Im}(\psi^* \nabla_k \psi) \tag{1.17}$$

associated with the wave function, Eq. (1.16) may also be written as

$$\frac{dQ(t)}{dt} = \frac{j^\psi}{|\psi|^2}(Q(t)) \tag{1.18}$$

Therefore, one can say that in the approach of Dürr, Goldstein, Tumulka and Zanghì, subatomic particles are assumed to have precise positions $\vec{Q}_k(t)$ at every time t that evolve according to Eq. (1.14) (or, the equivalent Eqs. (1.16) and (1.18)). Here, the Bohmian trajectories can be considered as the primitive ontology of the de Broglie-Bohm theory.

According to this re-reading of Bohmian Mechanics, the dynamics is completely defined by the Schrödinger equation and the guiding equation; there is no need of any further axioms involving the quantum potential which thus should not be regarded as the most fundamental structure of the theory. If the guiding Eq. (1.14)—or the equivalent Eqs. (1.16) and (1.18)—represents the basic equation of motion, Bohmian mechanics must be considered as a first-order theory, in which the fundamental quantity is the position of the particles, whose dynamics is directly and simply specified by the velocity field (1.15). This approach provides a first-order description of quantum processes, in the sense that, given the wavefunction, future and past motion may be calculated by simply specifying the positions of all particles under consideration at a given time. In this view, the second-order (Newtonian) concepts of acceleration, force, work and energy have not a fundamental role.

In this picture, moreover, contrary to what happens in the hydrodynamic one (where all the Bohmian trajectories associated with a given wave function—and corresponding to a different initial condition—are on an equal footing), only one of the Bohmian trajectories corresponds to reality, and all the other ones are no more than mathematical curves, representing possible alternative histories that could have occurred if the initial configuration had been different.

To synthesize, one can say that there is a wave-particle duality in a literal sense: there is a wave in R^{3N} and there are particles pursuing well-defined trajectories $\vec{Q}_1, \ldots, \vec{Q}_N$. The wave evolves according to the Schrödinger equation (1.13) and the particles move in a way that depends on the wave, namely according to the guiding equation, which can be considered as the most fundamental equation of motion in Bohmian Mechanics.

The quantum potential is assumed here to be a secondary physical entity with respect to the trajectories satisfying the guiding Eq. (1.14). The quantum potential emerges by taking the time derivative of the guidance Eq. (1.14) after some mathematical manipulations:

$$m_k \frac{d^2 \vec{Q}_k(t)}{dt^2} = -\nabla_k \left(V + U_{QP}^{\psi} \right) (Q(t)) \tag{1.19}$$

where

$$U_{QP}^{\psi} = -\sum_{j=1}^{N} \frac{\hbar^2}{2m_j} \frac{\nabla_j^2 |\psi|}{\psi} \tag{1.20}$$

is the quantum potential. While in the original Bohm's articles, Eqs. (1.3) (for one-body systems) and (1.8) (for many-body systems)—with the quantum potential given respectively by Eqs. (1.4) for one-body systems and (1.9) for many-body systems—were presented as the basic equations of motion and the guiding Eq. (1.14) can be seen as a constraint condition in order to exclude the solutions of (1.3) (or of (1.8)) which are not solutions of the guiding Eq. (1.14), in the view of Dürr, Goldstein, Tumulka and Zanghì the guiding Eq. (1.14) is considered as the only basic equation of motion. According to these authors, their approach presents an advantage with respect to the original Bohm's 1952 ideas in the sense that, since every solution of Eq. (1.14) is also a solution of Eq. (1.19), the Bohmian trajectories are the basic elements which allow us to solve the Schrödinger equation and thus to find the wave function responsible of the motion of the system under consideration. The second-order Eq. (1.19) serves only as a subsidiary equation which suggests how the Bohmian trajectories satisfying the fundamental guiding Eq. (1.14) can be used for solving the Schrödinger equation. By following [25], the algorithm is the following. Choose an ensemble of points (say $Q^{(1)}, \ldots, Q^{(n)}$ with very large n) in configuration space so that its distribution density is $\rho_0 = |\psi_0|^2$, ψ_0 being the initial function (which is supposed to be known). For each $Q^{(i)}$, determine the 3 N-vector of velocities $v^{\psi_0}(Q^{(i)})$ from the velocity law (1.15). For each i, solve the second-order equation of motion (1.19) with initial positions as in $Q^{(i)}$, initial velocities as in $v^{\psi_0}(Q^{(i)})$, and the quantum potential as defined in (1.20) but with $|\psi|$ replaced by $\sqrt{\rho}$, where ρ is the density in configuration space of the ensemble $Q^{(1)}, \ldots, Q^{(n)}$. This implies, for every time step $t \to t + \delta t$, to determine the quantum potential at time t from the density ρ of the ensemble points according to equation

$$U_{QP}^{\psi} = -\sum_{j=1}^{N} \frac{\hbar^2}{2m_j} \frac{\nabla_j^2 \sqrt{\rho}}{\sqrt{\rho}} \tag{1.21}$$

and to use Eq. (1.19) to propagate $Q^{(i)}(t)$ and $\frac{dQ^{(i)}}{dt}(t)$ by one time step and to obtain $Q^{(i)}(t + \delta t)$ and $\frac{dQ^{(i)}}{dt}(t + \delta t)$. The property of equivariance expressing the compatibility between the evolution of the wave function given by Schrödinger's equation and the evolution of the actual configuration given by the guiding Eq. (1.14) (which asserts that if $Q(0)$ is random with probability density given by $|\psi_0|^2$, then $Q(t)$ is also random, with probability density given by $|\psi|^2$), together with the fact that Eq. (1.19) follows from the Schrödinger equation (1.13) and the fundamental equation of motion (1.14), implies that this algorithm would yield exactly the family of Bohmian trajectories if we use infinite points with an initial distribution given exactly by $|\psi_0|^2$, if the time step δt is infinitesimal, and if no numerical error is involved in solving Eq. (1.19). With finite n and finite δt, we may obtain an approximation to the family of Bohmian trajectories. Once we have computed the trajectories, we can (more or less) obtain the wave function $\psi(q)$ up to a time-dependent phase factor as follows. If one writes the wave function as

$$\psi(q) = |\psi(q)|e^{iS(q)/\hbar} \tag{1.22}$$

then one obtains

$$\vec{v}_k^{\psi}(q) = \frac{1}{m_k} \nabla_k S. \tag{1.23}$$

Now, knowing all Bohmian trajectories then we can read off the 3 N-velocity $v(q,t) = (\vec{v}_1(q,t), \ldots, \vec{v}_N(q,t))$ of the trajectory that passes through $q \in R^{3N}$ at time t, solve

$$\vec{v}_k(q,t) = \frac{1}{m_k} \nabla_k S(q,t) \tag{1.24}$$

for $S(q,t)$, and finally set

$$\psi(q) = \sqrt{\rho(q,t)} e^{iS(q,t)/\hbar}. \tag{1.25}$$

Here one can observe that Eq. (1.24) determines the function $q \rightarrow S(q,t)$ up to a real constant time-dependent phase factor $\theta(t)$, that is to say, any $S(q,t) + \theta(t)$ is another solution of Eq. (1.24). As a consequence, $\psi(q,t)$ has been determined up to a phase factor $e^{i\theta(t)}$.

This algorithm implies that in the picture of Bohmian mechanics an ensemble of points in configuration space, with density ρ, leads to a quantum potential via Eq. (1.21) which is the physical cause of every trajectory but among these trajectories only one is real, all the others are mathematical curves.

Bohmian mechanics introduces the important question to explore in detail the ontological status of the wave-function appearing in the law of motion. As we have underlined before, in the approach of Dürr, Goldstein, Tumulka and Zanghì the primary physical reality is represented by the trajectories of the particles under consideration satisfying the guiding Eq. (1.14)—or the (1.16) or (1.18)—which is the fundamental law of motion, and, at the same time, there is a wave-particle duality in a literal sense and, according to the guiding equation, the wave function has the role to influence the motion of the particles (while the quantum potential can be considered as a secondary physical entity which emerges from the guiding equation and serves to determine the wave function responsible of the trajectories of the system under consideration). Thus, the following question becomes natural: must the wave function be considered as a physical entity having the same ontological status of the trajectories or not? In this regard, as already suggested by Dürr, Goldstein and Zanghì [21], Goldstein and Teufel [26] and Goldstein and Zanghì [27], and then showed in major detail by Esfeld, Lazarovici, Hubert and Dürr [28], the wave-function cannot be considered as a part of the fundamental ontology of Bohmian mechanics. According to these authors, the relationship between the wave-function and the motion of the particles inside the Bohmian mechanics is more appropriately conceived as a nomological one, instead of a causal one in terms of one physical entity acting on the other. In this picture, if one abandons the idea of the wave-function as a primary physical entity, the fundamental ontology consists only in the particles having their positions and their well-defined spatio-temporal trajectories. Thus, for instance, in the double slit experiment with one particle at a time, the particle goes through exactly one of the two slits, and that is all occurs in the physical world. According to the nomological interpretation of the wave function, there is no field or wave that guides the motion of the particle propagating through both slits so as to produce the interference pattern: the wave-function, as it appears in the law of motion, is merely a mathematical object defined on the configuration space, instead of a physical object existing in addition to the particles.

1.2.2 The Quasi-Newtonian Quantum Potential

In the fascinating article "A quasi-Newtonian approach to Bohmian quantum mechanics" [29], M. Atiq, M. Karamian and N. Golshani show that the Bohmian formulation of quantum mechanics can be obtained without the necessity to start from the Schrödinger equation. In the picture proposed by Atiq, Karamian and Golshani, the quantum potential can be considered as the basis of the Schrödinger equation rather than being a consequence of it, in other words can be considered as the fundamental concept, the fundamental physical reality responsible of quantum processes. These three authors also derived an equation that restricts the possible forms of quantum potential and determines the functional form of it without appealing to the wave function and the Schrödinger equation.

The quasi-Newtonian approach suggests to describe the quantum phenomena by the classical mechanics formalism. By considering the example of the double-slit interference, Atiq, Karamian and Golshani introduced an extra potential Q responsible for the quantum behaviours of the individual particles in this experiment. By denoting the density of the particles by ρ, and the mathematically unknown fundamental agent of these novel quantum behaviours by χ one has

$$\rho = f(\chi) \tag{1.26}$$

$$Q = Q(\chi). \tag{1.27}$$

By taking the inverse of Eq. (1.26)

$$\chi = f^{-1}(\rho) \tag{1.28}$$

and inserting this relation in Eq. (1.27) one can eliminate the unknown factor χ and obtain a direct dependence of Q on ρ:

$$Q = Q(\rho) \tag{1.29}$$

which, by introducing a function R such that

$$\rho = R^2 \tag{1.30}$$

can also be expressed as

$$Q = Q(R). \tag{1.31}$$

Equation (1.31) indicates that the extra potential Q could be function of R and· its partial derivatives. Now, one considers the Hamiltonian of a single particle in three dimensions:

$$H(x, p, t) = \frac{p^2}{2m} + V(x) + Q(R(x, t)) \tag{1.32}$$

where the quantum potential Q is taken to be the unknown function of R.

Here, in order to find the actual path of the particles, one needs extra assumptions about R and Q. It is here assumed that the total energy of the ensemble of particles must be minimized. This condition, according to variational calculus, leads to the following equation

$$\delta \int \rho (H - \lambda) d^3 x = 0 \tag{1.33}$$

in which λ is Lagrange's undetermined multiplier. In Eq. (1.33) the Hamiltonian can be expressed as a function of space coordinates only by using Hamilton-Jacobi's principal function S. According to the Hamilton-Jacobi theory, the momentum of particle is

$$p = \nabla S \tag{1.34}$$

and for conserved systems one has

$$S(x, t) = W(x) - Et \tag{1.35}$$

and hence Eq. (1.33) may be rewritten as

$$\delta \int R^2 \left\{ \frac{(\nabla S)^2}{2m} + V + Q - \lambda \right\} d^3 x = 0. \tag{1.36}$$

Equation (1.36), if the quantum potential Q is in the Bohmian form, turns out to be an integral form of the time-independent Schrödinger equation with λ as the eigenvalue. In fact, for stationary states using the quantum potential (1.4), Eq. (1.36) becomes

$$\delta \int \{ R^2 H - \lambda R^2 \} d^3 x = \delta \int \psi^* \{ \hat{H}\psi - \lambda \psi \} d^3 x = 0 \tag{1.37}$$

and variation with respect to ψ^* leads to the time-independent Schrödinger equation with eigenvalue λ

$$\hat{H}\psi = \lambda \psi. \tag{1.38}$$

Now, if one denotes the integrand of (1.37) by g, using summation rule for indices i, j and abbreviation ∂_i for the partial derivative $\frac{\partial}{\partial x_i}$, and so on, from variational calculus one obtains

$$\frac{\partial g}{\partial R} - \partial_i\left(\frac{\partial g}{\partial(\partial_i R)}\right) + \partial_i\partial_j\left(\frac{\partial g}{\partial(\partial_i\partial_j R)}\right) = 0 \qquad (1.39)$$

$$\partial_i\left(\frac{\partial g}{\partial(\partial_i S)}\right) = 0. \qquad (1.40)$$

From Eq. (1.39) it follows that:

$$2R\left\{\frac{(\nabla S)^2}{2m} + V + Q - \lambda\right\} + R^2\frac{\partial Q}{\partial R} - \partial_i\left(R^2\frac{\partial Q}{\partial(\partial_i R)}\right) + \partial_i\partial_j\left(R^2\frac{\partial Q}{\partial(\partial_i\partial_j R)}\right) = 0 \qquad (1.41)$$

namely, taking into account that $\left\{\frac{(\nabla S)^2}{2m} + V + Q - \lambda\right\} = E - \lambda$, one obtains

$$R^2\frac{\partial Q}{\partial R} - \partial_i\left(R^2\frac{\partial Q}{\partial(\partial_i R)}\right) + \partial_i\partial_j\left(R^2\frac{\partial Q}{\partial(\partial_i\partial_j R)}\right) = 2R(\lambda - E). \qquad (1.42)$$

Since the particle energy E and the energy eigenvalue λ are both constants, they are taken to be identical. Therefore, one obtains the following relations

$$\lambda = E = \frac{(\nabla S)^2}{2m} + V(x) + Q \qquad (1.43)$$

$$R^2\frac{\partial Q}{\partial R} - \partial_i\left(R^2\frac{\partial Q}{\partial(\partial_i R)}\right) + \partial_i\partial_j\left(R^2\frac{\partial Q}{\partial(\partial_i\partial_j R)}\right) = 0. \qquad (1.44)$$

In this way, one finds that this extra potential Q responsible for quantum phenomena obtained by minimizing the total energy of the ensemble of particles, the quantum potential, must satisfy the fundamental Eq. (1.44). Moreover, from Eq. (1.40) one gets

$$\nabla \cdot \left(R^2\frac{\nabla S}{m}\right) = 0 \qquad (1.45)$$

which is the continuity equation for stationary states.

The mathematical expression of Q in function of R must be such that the Eq. (1.44) is satisfied for every arbitrary R. Atiq, Karamian and Golshani showed that the first-order derivatives are not sufficient for getting a non-trivial quantum potential. Since the quantum potential Q is a scalar function and therefore must be rotational-invariant, the first and second derivatives of R must appear in the form of $|\nabla R|$ and $\nabla^2 R$, respectively. So, Atiq, Karamian and Golshani demonstrated that non-trivial solutions of Eq. (1.44) have the form

$$Q = AR^m |\nabla R|^n (\nabla^2 R)^p \tag{1.46}$$

where the powers m, n and p can assume only the two following sets of values:

$$\text{m} = 0, \ \text{n} = 0, \ \text{p} = 0 \tag{1.47}$$

$$\text{m} = -1, \ \text{n} = 0, \ \text{p} = 1. \tag{1.48}$$

In the case (1.47), one obtains the trivial solution $Q(x) = A = const$. The second case leads to the result that one has in ordinary Bohmian mechanics, namely

$$Q(x) = A \frac{\nabla^2 R}{R}. \tag{1.49}$$

Therefore, not only the quantum Hamilton-Jacobi equation and the continuity equation are obtained, but the form of the quantum potential Q in terms of R is justified. In quasi-Newtonian theory, the form of quantum potential (1.49) is a mathematical result which derives from minimizing the total energy of the ensemble rather than being a direct consequence of the Schrödinger equation.

Moreover, Atiq, Karamian and Golshani underline that the constant value of A and specifically its sign in the Eq. (1.49) are significant. In particular, in this approach the value of the constant $A = -\frac{\hbar^2}{2m}$ emerges directly, by adapting the energy levels of Hydrogen atom in the theory with those obtained from Bohr's model (in a similar way to what Schrödinger did in his original works for finding some constants [30]).

It is important to emphasize that in this treatment, the concept of wave function—intended as primary physical entity—is not used. In this approach, S is a mathematical function, whose derivatives provide momentum and energy, and R, is representative of a new physical entity which contributes to the quantum dynamics of the particle through the extra potential—namely the quantum potential—Q. This means that here the quantum potential can be considered as a more fundamental concept than wave function and Schrödinger equation.

If the state is not stationary, i.e., R depends on time and S is a general function of time and space, by using the following equation

$$\delta \int R^2 \left\{ H(x, S(x,t), R(x,t)) + \frac{\partial S}{\partial t} \right\} d^4x = 0 \tag{1.50}$$

and the Eqs. (1.39), (1.40) and (1.49), one gets the Hamilton-Jacobi and continuity equations:

$$\frac{(\nabla S)^2}{2m} + V(x) + Q + \frac{\partial S}{\partial t} = 0 \tag{1.51}$$

$$\frac{\partial R^2}{\partial t} + \nabla \cdot \left(R^2 \frac{\nabla S}{m} \right) = 0. \tag{1.52}$$

Mathematically, the set of Eqs. (1.51) and (1.52) along with the condition

$$\oint \nabla S \cdot dx = nh \tag{1.53}$$

(which means that the phase of wave function is unique and thus the wave function is single-valued) is equivalent to the Schrödinger equation for $\psi = R \, e^{iS/\hbar}$. Therefore if one imposes the extra uniqueness condition on S (which is nevertheless not necessary here), one can obtain the Schrödinger equation. Establishing the condition (1.53) is equivalent to fix the wave function. In Atiq's, Karamian's and Golshani's approach, there is no need and no reason for imposing the condition (1.53), this means that one does not appeal to the wave function. This can be considered one of the most relevant differences between the quasi-Newtonian approach and the Bohmian mechanics: indeed, in the quasi-Newtonian approach S is the same as classical Hamilton-Jacobi principal function, there is no uniqueness condition (1.53) on S. Another significant difference between quasi-Newtonian approach and the Bohmian formulation is connected with the denial of the condition (1.53). When we do not require to consider the function S as phase of a wave function there is no need to consider R as amplitude of a wave function. Therefore, there is no need to consider R as a positive-definite function. Bohmian mechanics and also the approach here examined imply that R appears in the forms of R^2 or $\frac{\nabla^2 R}{R}$, thus the negative values for R are not a problem. The imposition of the uniqueness condition on S and the positive-definiteness condition on R are not needed in order to describe quantum phenomena.

The importance of the quantum potential as a fundamental entity responsible for quantum phenomena (that must be added in a quasi-Newtonian picture in order to reproduce the quantum processes) is linked also to the fact that, in the context of the non-relativistic de Broglie-Bohm theory, a Bohmian path integral—depending on the combined action of the classical potential and the quantum potential—can be associated with the Bohmian trajectory of the particle. In this regard, in the paper *The path integral approach in the frame work of causal interpretation* M. Abolhasani and M. Golshani [31] showed that the propagation of the wave function in the context of de Broglie-Bohm theory for the one-body system can be achieved by means of a Bohmian path integral which, for two points $(x; t)$ and $(x_0; t_0)$ with a finite distance on a Bohmian path, is defined by the relation

$$\psi(\vec{x}, t) = \exp\left\{ \frac{i}{\hbar} \int_{\vec{x}_0, t_0}^{\vec{x}, t} \left[\frac{(\nabla S)^2}{2m} - (Q + V) \right] dt - \int_{\vec{x}_0, t_0}^{\vec{x}, t} \frac{\nabla^2 S}{2m} dt \right\} \psi(\vec{x}_0, t_0) \tag{1.54}$$

where the first exponential can be obtained by integrating the quantum Hamilton-Jacobi equation (1.3) on the Bohmian path while the second exponential can be obtained by integrating the continuity Eq. (1.6). Equation (1.54) shows that the classical action (described by the Feynman path integral) is replaced by the quantum action (which is linked with the quantum potential Q given by Eq. (1.4)). In analogous way, for a system of N particles the Bohmian path integral can be expressed in the following way

$$
\psi(\vec{x}_1,\ldots,\vec{x}_N,t) = \exp\left\{ \frac{i}{\hbar} \int\limits_{\vec{x}_0,\,t_0}^{\vec{x},\,t} \sum_{i=1}^{N} \left[\frac{(\nabla_i S)^2}{2m_i} - (Q+V) \right] dt - \int\limits_{\vec{x}_0,\,t_0}^{\vec{x},\,t} \sum_{i=1}^{N} \frac{\nabla_i^2 S}{2m_i} dt \right\}
$$
$$
\psi(\vec{x}_{01},\ldots,\vec{x}_{0N},t_0)
$$

$$(1.55)$$

with the quantum potential Q given by Eq. (1.9). In particular, Abolhasani and Golshani determined $\psi(\vec{x},t)$ for a free-wave packet in one dimension in terms of its Fourier components e^{ikx} (for which $\nabla^2 S = 0$ and $Q = 0$). In this case the Bohmian path integral (1.54) becomes

$$
\psi(x,t) = \int dk e^{-i\frac{\hbar k^2}{2m}(t-t_0)} \phi(k) e^{ikx}
$$

$$(1.56)$$

the Fourier-Bohm path integrals. Abolhasani's and Golshani's treatment shows furthermore that for the free wave packet in one dimension the Feynman path integral can be obtained directly from the Bohmian path integral (1.54) on the basis of an "heuristic argument" (as a consequence of the fact that in this simple case the quantum potential vanishes). The trajectory of a physical system corresponding to a Bohmian path integral (Eq. (1.54) for one-body systems and Eq. (1.55) for many-body systems) can be seen as the effect of the action of a quantum potential emerging as an extra term which reproduces the novel features of quantum phenomena (with respect to the classical situation) in the context of a quasi-Newtonian approach.

1.2.3 The Thermodynamic Way to the Quantum Potential

In the recent articles *The vacuum fluctuation theorem: Exact Schrödinger equation via non-equilibrium thermodynamics* [32] and *On the thermodynamic origin of the quantum potential* [33] G. Grössing proposed the idea that quantum phenomena have a basic thermodynamic origin and, in particular, that the quantum potential arises from the presence of a subtle thermal vacuum energy distributed across the whole domain of an experimental arrangement. In this interpretation, the form of the quantum potential is demonstrated to be exactly identical to the heat distribution derived from the defining equation for classical diffusion-wave fields.

First of all, in the paper [32], Grössing suggests to merge some results of non-equilibrium thermodynamics with classical wave mechanics in such a way that the many microscopic degrees of freedom associated with a subquantum medium can be embedded into the more "macroscopic" properties which characterize the wave-like behaviour on the quantum level. By considering a particle immersed in a "heat bath", namely a reservoir that is very large compared to the small dissipative system, in the picture of a Maxwell-Boltzmann distribution for the momentum, one arrives at the equilibrium-type probability (density) ratio

$$\frac{P(\vec{x}, t)}{P(\vec{x}, 0)} = e^{-\frac{\Delta Q}{kT}} \tag{1.57}$$

with k being Boltzmann's constant, T the reservoir temperature and ΔQ the heat that is exchanged between the particle and its environment. In Grössing's view, Eq. (1.57), together with the assumption that quantum particles are actually dissipative systems maintained in a non-equilibrium steady-state by a permanent throughput of energy or heat flow, respectively (which means that the total energy of the particle is given by equation

$$E_{tot} = \hbar\omega + \frac{(\delta p)^2}{2m} \tag{1.58}$$

where δp is the additional, fluctuating momentum component of the particle), and that the detection probability density provided by the particle's environment is considered to coincide with a classical wave's intensity according to relation

$$P(\vec{x}, t) = R^2(\vec{x}, t) \tag{1.59}$$

with $R(x, t)$ being the wave's real-valued amplitude, allows us to obtain Schrödinger's equation from classical mechanics with only two supplementary well-known observations. The first is given by the following relation between heat ΔQ and action S

$$\Delta Q = 2\omega[\delta S(t) - \delta S(0)] \tag{1.60}$$

the second consists in the requirement that the average kinetic energy of the thermostat equals the average kinetic energy of the oscillator, for each degree of freedom, as

$$\frac{kT}{2} = \frac{\hbar\omega}{2} \tag{1.61}$$

Combining Eqs. (1.57), (1.60) and (1.61), one obtains

$$P(\vec{x}, t) = P(\vec{x}, 0)e^{-\frac{2}{\hbar}[\delta S(\vec{x}, t) - \delta S(\vec{x}, 0)]} \tag{1.62}$$

and thus the expression of the momentum fluctuation $\delta\vec{p}$ in (1.58) is

$$\delta P(\vec{x}, t) = -\frac{\hbar}{2}\frac{\nabla P(\vec{x}, t)}{P(\vec{x}, t)} \tag{1.63}$$

As a consequence, the action integral for a system of n particles turns out to be

$$A = \int P \left[\frac{\partial S}{\partial t} + \sum_{i=1}^{n} \frac{1}{2m_i} \nabla_i S \nabla_i S + \sum_{i=1}^{n} \frac{1}{2m_i} \left(\frac{\hbar}{2} \frac{\nabla_i P}{P} \right)^2 + V \right] d^n x dt \qquad (1.64)$$

where $P = P(\vec{x}_1, \vec{x}_2, \ldots, \vec{x}_n, t)$. Here, introducing the Madelung transformation

$$\psi = R \, e^{iS/\hbar} \qquad (1.65)$$

where $R = \sqrt{P}$ as in (1.59), one obtains

$$A = \int d^n x dt \left[|\psi|^2 \left(\frac{\partial S}{\partial t} + V \right) + \sum_{i=1}^{n} \frac{\hbar^2}{2m_i} |\nabla_i \psi|^2 \right] \qquad (1.66)$$

The action (1.66), by using the identity $|\psi|^2 \frac{\partial S}{\partial t} = -\frac{i\hbar}{2} (\psi^* \dot{\psi} - \dot{\psi}\psi^*)$, leads to the n-particle Schrödinger equation

$$i\hbar \frac{\partial \psi}{\partial t} = \left(-\sum_{i=1}^{n} \frac{\hbar^2}{2m_i} \nabla_i^2 + V \right) \psi \qquad (1.67)$$

Equation (1.64), if one applies a variation in P, leads to the quantum Hamilton-Jacobi equation of the de Broglie-Bohm theory, i.e.,

$$\frac{\partial S}{\partial t} + \sum_{i=1}^{n} \frac{(\nabla_i S)^2}{2m_i} + V(\vec{x}_1, \ldots, \vec{x}_n, t) + U_q(\vec{x}_1, \ldots, \vec{x}_n, t) = 0 \qquad (1.68)$$

where U_q is the "quantum potential"

$$U_q(\vec{x}_1, \ldots, \vec{x}_n, t) = \sum_{i=1}^{n} \frac{\hbar^2}{4m_i} \left[\frac{1}{2} \left(\frac{\nabla_i P}{P} \right)^2 - \frac{\nabla_i^2 P}{P} \right] = -\sum_{i=1}^{n} \frac{\hbar^2}{2m_i} \frac{\nabla_i^2 R}{R} \qquad (1.69)$$

Moreover, with the definitions

$$\vec{u}_i = \frac{\delta p_i}{m_i} = -\frac{\hbar}{2m_i} \frac{\nabla_i P}{P} \quad \text{and} \quad \vec{k}_{\bar{u}_i} = -\frac{1}{2} \frac{\nabla_i P}{P} = -\frac{\nabla_i R}{R} \qquad (1.70)$$

one can express the quantum potential (1.69) as

$$U_q(\vec{x}_1, \ldots, \vec{x}_n, t) = \sum_{i=1}^{n} \left[\frac{m_i \vec{u}_i \cdot \vec{u}_i}{2} - \frac{\hbar}{2} (\nabla_i \cdot \vec{u}_i) \right] = \sum_{i=1}^{n} \left[\frac{\hbar^2}{2m_i} \left(\vec{k}_{\bar{u}_i} \cdot \vec{k}_{\bar{u}_i} - \nabla_i \cdot \vec{k}_{\bar{u}_i} \right) \right]$$

$$(1.71)$$

and inserting the dependence of $\vec{u}_i$ on the spatial behaviour of the heat flow expressed by relation

$$\vec{u}_i = \frac{1}{2\omega_i m_i} \nabla_i Q \qquad (1.72)$$

one obtains the thermodynamic formulation of the quantum potential as

$$U_q(\vec{x}_1, \ldots, \vec{x}_n, t) = \sum_{i=1}^{n} \frac{\hbar^2}{4m_i} \left[\frac{1}{2} \left(\frac{\nabla_i Q}{\hbar \omega_i} \right)^2 - \frac{\nabla_i^2 Q}{\hbar \omega_i} \right]. \tag{1.73}$$

On the basis of treatment of Eqs. (1.59), (1.70), (1.72) and (1.73), the following interpretation of the quantum potential becomes possible: the detection probability density provided by the particle's environment (and which coincides with a classical wave's intensity) determines a heat flow having a spatial behaviour given by Eq. (1.72), and this heat flow determines the quantum potential (1.73) responsible for the quantum behaviour of the particle. According to Eq. (1.73), the heat flow between the particle under consideration and its environment can be considered as the source of the quantum potential of non-relativistic de Broglie-Bohm theory.

This view implies that the energetic scenario of a steady-state oscillator in non-equilibrium thermodynamics is associated with a throughput of heat, i.e., a kinetic energy at the subquantum level which can achieve the two following aims: to provide the necessary energy to maintain a constant oscillation frequency ω and some excess kinetic energy which results in a fluctuating momentum contribution $\delta\vec{p}$ to the momentum $\vec{p}$ of the particle. Moreover, the "particle" in this thermo-dynamic environment will not only receive kinetic energy from it, but, in order to balance the stochastic influence of the additional momentum fluctuations, will also dissipate heat into the environment. In fact, on the basis of the "vacuum fluctuation theorem" introduced in the paper [32], the larger the energy fluctuation of the oscillating system is, the higher is the probability that heat will be dissipated into the environment rather than to be absorbed. Thus, if one considers the stochastic "forward" movement, namely $\delta\vec{p}/m = \vec{u}$, or the current $\vec{J} = P\vec{u}$, respectively, these are balanced by the osmotic velocity $-\vec{u}$, or $\vec{J} = -P\vec{u}$, respectively.

Inserting Eq. (1.70) into the definition of the "forward" diffusive current $\vec{J}$ one has

$$\vec{J} = P\vec{u} = -\frac{\hbar}{2m} \nabla P \tag{1.74}$$

which, if one takes account of the continuity equation $\dot{P} = -\nabla \cdot \vec{J}$, becomes

$$\frac{\partial P}{\partial t} = \frac{\hbar}{2m} \nabla^2 P. \tag{1.75}$$

Equations (1.74) and (1.75) are the first and second of Fick's laws of diffusion, respectively (if one introduces the diffusivity $D = \frac{\hbar}{2m}$), and $\vec{J}$ is the diffusion current.

In the paper [33] Grössing analysed the perspectives introduced by the "osmotic" type of dissipation of energy from the particle to its environment in the interpretation of the quantum potential. In this regard, in virtue of Eq. (1.58) and the strict directionality of any heat flow, one can redefine Eq. (1.72) for the case of heat dissipation where $\Delta Q = Q(t) - Q(0) < 0$. If one assumes that the heat flow is a positive quantity, i.e. it provides a measure of the positive amount of heat dissipated into the environment, one can choose the negative of the above expression, $-\Delta Q$, and insert this into Eq. (1.72), in order to obtain the osmotic velocity

$$\vec{u} = D\frac{\nabla P}{P} = -\frac{1}{2\omega m}\nabla Q \tag{1.76}$$

and thus the corresponding osmotic current is

$$\vec{J} = P\vec{u} = D\nabla P = -\frac{P}{2\omega m}\nabla Q. \tag{1.77}$$

Then, from the corollary to Fick's second law, one obtains

$$\frac{\partial P}{\partial t} = -\nabla \cdot \vec{J} = -D\nabla^2 P = \frac{1}{2\omega m}\left[\nabla P \cdot \nabla Q + P\nabla^2 Q\right]. \tag{1.78}$$

Equations 1.74–1.78 allow us to understand the thermodynamic meaning of the quantum potential expressed by Eq. (1.73). In this regard, for simplicity we base our discussion on one-body systems.

Let us now consider the simplest case $U_q = 0$. In this case, from Eq. (1.73) the thermodynamic corollary of a vanishing quantum potential for the one-body system is:

$$\nabla^2 Q = \frac{1}{2\hbar\omega}(\nabla Q)^2. \tag{1.79}$$

The osmotic flux conservation (1.78), by using Eq. (1.57) and (1.61) and (1.79), leads thus to the following relation

$$\frac{1}{2}\left(\frac{\nabla Q}{\hbar\omega}\right)^2 = \frac{1}{D}\frac{1}{\hbar\omega}\frac{\partial Q}{\partial t} \tag{1.80}$$

Now inserting Eq. (1.80) into Eq. (1.73) one gets, with $\hbar\omega = $ constant and $\hat{Q} = Q/\hbar\omega$,

$$U_q = -\frac{\hbar^2}{4m}\left[\nabla^2\hat{Q} - \frac{1}{D}\frac{\partial\hat{Q}}{\partial t}\right] \tag{1.81}$$

or, more in general,

$$U_q = -\frac{\hbar^2}{4m}\left[\nabla^2 Q - \frac{1}{D}\frac{\partial Q}{\partial t}\right]. \tag{1.82}$$

In the case $U_q = 0$, one obtains

$$\left[\nabla^2 Q - \frac{1}{D}\frac{\partial Q}{\partial t}\right] = 0 \tag{1.83}$$

which expresses the classical heat conduction equation. In other words, even for free particles, both in the classical and in the quantum case, one can identify a heat dissipation process emanating from the particle. Therefore, in Grössing's approach, a non-vanishing "quantum potential" is a way to describe the spatial and temporal correlations of the corresponding thermal flow in case the particle is not free.

In the general case $U_q \neq 0$ one can introduce an explicitly non-vanishing source term for the quantum potential (1.82), i.e.,

$$\left[\nabla^2 Q - \frac{1}{D}\frac{\partial Q}{\partial t}\right] = q(x)e^{i\omega t} \neq 0 \tag{1.84}$$

which can be solved via separation of variables. Thus, with the *ansatz*

$$Q = X(x)T(t), \text{ with } T = e^{i\omega t} \tag{1.85}$$

and

$$q(x) = \alpha(x)X \tag{1.86}$$

division by (XT) then provides the constant

$$\frac{\nabla^2 X}{X} = \frac{\frac{\partial}{\partial t}T}{DT} + \alpha = -\lambda. \tag{1.87}$$

Considering, for example, the case of a particle entrapped in a box of length L whose walls are infinitely high, one can introduce the Dirichlet boundary conditions, namely

$$Q(0,t) = Q(L,t) = 0 \tag{1.88}$$

which provides the constant λ as

$$\lambda = \frac{n^2\pi^2}{L^2} \equiv k_n^2. \tag{1.89}$$

Equation (1.87) leads to the solution

$$X = NC_Q \sin\left(\frac{n\pi}{L}x\right) \tag{1.90}$$

where N is a normalization constant and C_Q is a dimensionality preserving constant. Furthermore, with $T = e^{i\omega_n t}$ one has $\frac{i\omega_n}{D} + \alpha = -k_n^2$, and therefore

$$\alpha = -k_n^2(1+i). \tag{1.91}$$

Equation (1.86) leads thus to the following solution for $q(x)$:

$$q(x) = -k_n^2(1 + i)NC_Q \sin\left(\frac{n\pi}{L}x\right) \tag{1.92}$$

and, with (1.85), (1.89), and $\tilde{Q} = Q/C_Q$,

$$\tilde{Q}(x, t) = N \sin(k_n x)e^{i\omega_n t}, \tag{1.93}$$

In virtue of the Dirichlet boundary conditions, the eigenvalue equation of the laplacian

$$\nabla^2 e_n = -k_n^2 e_n \tag{1.94}$$

has solutions $e_n = N \sin(k_n x)$
   where

$$\langle e_n \mid e_m \rangle = \int e_n(x)e_m(x)dx = \begin{cases} 0, m \neq n \\ 1, m = n \end{cases}. \tag{1.95}$$

This means that, for $m = n$, (1.95) can be interpreted as a probability density, with

$$\int\limits_0^L P dx = N^2 \int\limits_0^L \sin^2(k_n x)dx = 1 \tag{1.96}$$

From Eq. (1.96) follows

$$N = \sqrt{\frac{2}{L}}. \tag{1.97}$$

Therefore, the heat distribution in the box is given by

$$\tilde{Q}(x, t) = \sqrt{\frac{2}{L}} \sin(k_n x)e^{i\omega_n t} \tag{1.98}$$

with the probability density

$$P = \left|\tilde{Q}(x, t)\right|^2 \tag{1.99}$$

As a consequence, the classical state (1.98) turns out to be identical to the quantum mechanical one, and Eqs. (1.99) and (1.69) lead to a quantum potential given by the following relation

$$U_q = \frac{\hbar^2 k_n^2}{2m}. \tag{1.100}$$

The interpretation suggested by Grössing shows a strict link between the quantum potential with diffusion-wave fields. In fact, introducing the solutions (1.92) into Eq. (1.84), one obtains a sort of "eigenvalue equation" of the form

$$\left[\nabla^2\tilde{Q} - \frac{1}{D}\frac{\partial\tilde{Q}}{\partial t}\right] = q(x)e^{i\omega t} = -(1+i)k_n^2\tilde{Q} \tag{1.101}$$

which, applying a temporal Fourier transformation and introducing the complex diffusion number $\kappa(x,\omega) \equiv \sqrt{\frac{i\omega}{D}}$, becomes a Helmholtz-type pseudo-wave equation

$$\nabla^2\tilde{Q}(x,\omega) - \kappa^2(x,\omega) = \sigma(x,\omega). \tag{1.102}$$

Equation (1.146) is exactly the defining equation for a thermal-wave-field and therefore describes the spatio-temporal behaviour of diffusion waves.

Moreover, in [33] Grössing showed that by taking into account both the osmotic current of heat dissipation and the usual forward current, the total quantum potential, integrated over periods of time $t \approx n/\omega$ long enough so that it is characterized by the energy throughput of n total currents, or by equal weights of n "absorption" and n "dissipation" currents, respectively, is—in virtue of equation (1.82)—given by the following relation

$$\bar{U}_q = \frac{1}{n}\left(n\bar{U}_{forward} + n\bar{U}_{osmotic}\right) = -\frac{\hbar^2}{4m}\frac{1}{\hbar\omega}\left[2\nabla^2 Q + \frac{1}{D}\frac{\partial Q}{\partial t} - \frac{1}{D}\frac{\partial Q}{\partial t}\right]$$
$$= -\frac{\hbar^2}{2m}\frac{\nabla^2 Q}{\hbar\omega}. \tag{1.103}$$

In the thermalized version of the quantum potential, given by Eq. (1.103), the usual quantum Hamilton-Jacobi equation (1.68) of de Broglie-Bohm theory can be rewritten as

$$\frac{\partial S}{\partial t} + \sum_{i=1}^{n}\frac{(\nabla_i S)^2}{2m_i} + V - \sum_{i=1}^{n}\left(\frac{L(\omega_i)}{2}\right)^2\nabla_i^2 Q = 0 \tag{1.104}$$

Thus, the equations of quantum motion can be written as

$$m_i\frac{d\vec{v}_i}{dt} = -\nabla_i(V + U_q) = -\nabla_i V + \sum_{i=1}^{n}\left(\frac{L(\omega_i)}{2}\right)^2\nabla_i(\nabla_i^2 Q) = 0 \tag{1.105}$$

which, at least for simple solutions Q, allows a simplification in the calculation of quantum trajectories.

Therefore, in Grössing's approach, the "form" of the quantum potential, as given by $-\frac{\nabla^2 R}{R}$, is linked essentially to a Helmholtz-type dependence $-\nabla^2 Q$ of a thermal energy Q which is distributed "non-locally" throughout an experimental apparatus, for example. A fundamental feature of diffusion-wave fields is represented by non-locality. The "propagation speed" of these fields is infinite and, thus, if one considers a prepared neutron source in a reactor, one immediately has a thermal field in the "vacuum" that non-locally links the neutron oven, the apparatus (including, for example, a Mach-Zender interferometer), and the detectors.

The (typical) Gaussians used to describe the initial quantum mechanical particle distributions thus also contribute to the form of the heat distribution in the overall system, no matter which particle actually is on its way through the interferometer. In this way, all "potential" paths are implicitly present throughout the experiment (i.e., under constant boundary conditions) in that the corresponding thermal field exists no matter where the particle actually is. On the basis of the infinite propagation of the diffusion-wave fields, one obtains in this way a promising perspective for a deeper understanding of the physical origin of quantum non-locality characterizing Bohm's quantum potential.

The wave equation corresponding to the Helmholtz-type dependence $-\nabla^2 Q$ is a parabolic one and thus one must generally expect the thermal waves' behaviour to be radically different from that of ordinary hyperbolic ones. In fact, Grössing have discussed several examples of forms of the quantum potential which clearly exhibit the departure from ordinary wave behaviour in terms of the appearance of accumulation and depletion zones. Thus, the "strange" form of the quantum potential as well as the (only seemingly "surrealistic") Bohmian trajectories characterizing classic quantum phenomena like the double-slit interference can potentially be fully understood with the aid of the physics of diffusion-wave fields. Identifying the quantum potential appearing in the quantum Hamilton-Jacobi equation of the de Broglie-Bohm theory with the presence of diffusion-wave fields, in several classic quantum phenomena the Bohmian trajectories, deriving from Bohm's quantum potential, can be understood in terms of the presence of diffusion-wave fields.

For example, let us consider the double-slit interference. In this case, the experimental results are explained through a quantum potential characterized by the appearance of "stripes", or "deep valleys" and "high-level plateaus", respectively and the corresponding trajectories indicate that the particles are accelerated in the "valleys" and move unhindered on the plateaus. Moreover, the particle trajectories from each side are "bouncing off" a central symmetry line such that no particle from the left slit can enter the right region, and vice versa. This bouncing off could be criticized as "unphysical" because apparently the momentum conservation would be severely violated on a more macroscopic scale (namely taking account of the positions of the two slits). However, in Grössing's thermodynamic interpretation of the quantum potential, in virtue of the presence of diffusion-wave fields, an overall momentum conservation is given and thus that criticism completely dissolves. Both the "mountains" between the two slit regions and the "stripes" of alternating valleys and plateaus, respectively, are the expression of a structured reservoir of heat (kinetic energy) due to the presence of accumulation and depletion zones of the non-local diffusion-wave fields. The particles emerging from the slits are supposed to emit spherically symmetrical thermal waves from the origin of their respective positions and, after completion of the interaction to the whole thermal field, the "heat" will dissipate far away from the slit system, making the quantum potential effectively flat on its outer edges. In other words, in the central region the kinetic energy reservoir between the two slits (the "mountain") will produce the heat waves which will strongly change the momentum between the particles radiating them away from the axis of the central symmetry. In this way, the emerging "stripes" represent indeed accumulation and depletion zones of the diffusion-wave fields. In particular, here the quantum potential represents the thermal wave field in its totality, as the latter is already there across the whole experimental setup, namely, due to the infinitely fast propagation of the diffusion-wave fields. This explains the peculiar form of the quantum potential as well as the unusual behaviour of particle motion.

1.2.4 The Geometrodynamic Approach to the Quantum Potential

On the basis of Bohm's and Hiley's 1984 work, the crucial feature of the quantum potential lies in its active information, a non-local global information on the environment in which the experiment is performed. This means that the information determined by the quantum potential can be defined as a geometric information "woven" into space-time. Quantum potential has a geometric contextual nature and at the same time is a dynamical entity just because its information about the process and the environment is active, determines the behaviour of the quantum objects (for the geometrodynamic picture see Fiscaletti [34]).

According to the research of several authors, there is the possibility to understand and describe quantum mechanics on the basis of a geometric interpretation. In this context, the quantum potential may be seen as the information channel expressing the modification of the geometrical properties of the configuration space in the quantum domain. In particular, Weyl geometries introduce interesting perspectives towards a new reading of quantum mechanics and thus of the quantum potential. Instead of imposing a priori that quantum mechanics must be constructed over an Euclidean background, quantum processes can be seen as a manifestation of a non-Euclidean structure derived from a variational principle: Bohm's quantum potential can be seen as the fundamental geometrodynamic entity which expresses the deformation of the geometrical properties of the configuration space of the physical processes with respect to the Euclidean space of classical physics. In this alternative picture, the validity of the specific geometrical structure proposed can be checked a posteriori comparing it to the usual non-relativistic quantum mechanics.

As regards the interpretation of the quantum potential in a geometrodynamic picture, in the recent article *On a geometrical description of quantum mechanics*, M. Novello, J. M. Salim and F. Falciano showed that there is a tight connection between Bohm's quantum potential and the Weyl integrable space [35]. They found out that Bohm's quantum potential can be identified with the scalar curvature of the Weyl integrable space and developed a variational principle that reproduces the Bohmian equations of motion. In synthesis, Novello's, Salim's and

Falciano's model suggests that one can reinterpret quantum mechanics as a manifestation of non-Euclidean structure of the three-dimensional space: by starting from a variational principle which defines the non-Euclidean structure of space, they provided a geometrical interpretation for quantum processes in the picture of the Weyl geometry.

In Novello's, Salim's and Falciano's approach, the geometrical structure of space is obtained by starting from the action

$$I = \int dt d^3x \sqrt{g} \Omega^2 \left(\lambda^2 R - \frac{\partial S}{\partial t} - H_m \right) \qquad (1.106)$$

where $g = \det g_{ij}$, $R \equiv g^{ij} R_{ij}$, $R_{ij} = \Gamma^m_{mi,j} - \Gamma^m_{ij,m} + \Gamma^l_{mi}\Gamma^m_{jl} - \Gamma^l_{ij}\Gamma^m_{lm}$ is the Ricci curvature tensor, λ^2 is a constant having dimension of energy multiplied for length squared. The connection of the 3D-space Γ^i_{jk}, the Hamilton's principal function S and the scalar function Ω are the independent variables of the approach.

In the case of a point-like particle the matter Hamiltonian is $H_m = \frac{1}{2m} \nabla S \cdot \nabla S + V$. Variation of the action (1.106) with respect to the independent variables gives

$$g_{ij;k} = -4(\ln \Omega)_{,k} g_{ij} \qquad (1.107)$$

where ";" denotes covariant derivative and "," simple spatial derivative. Variation with respect to Ω gives

$$\lambda^2 R = \frac{\partial S}{\partial t} + \frac{1}{2m} \nabla S \cdot \nabla S + V. \qquad (1.108)$$

Equation (1.108) describes the affine properties of the physical space. On the other hand, by setting $\lambda^2 = \frac{\hbar^2}{16m}$ and taking into account the expression of the curvature in terms of the scalar function Ω, $R = 8 \frac{\nabla^2 \Omega}{\Omega}$, Eq. (1.108) becomes

$$\frac{\partial S}{\partial t} + \frac{1}{2m} \nabla S \cdot \nabla S + V - \frac{\hbar^2}{2m} \frac{\nabla^2 \Omega}{\Omega} = 0. \qquad (1.109)$$

Equation (1.109) is analogous to the quantum Hamilton-Jacobi equation (1.3) of de Broglie-Bohm interpretation of quantum mechanics, if one identifies the scalar function Ω with the amplitude of the wave function. In the approach based on Weyl integrable space, the quantum potential can be thus identified with the scalar curvature which characterizes this geometry.

The inverse square root of the scalar curvature defines a typical length (Weyl length) that can be used to evaluate the strength of quantum effects

$$L_W = \frac{1}{\sqrt{R}}. \qquad (1.110)$$

Inside this picture, the classical behaviour is obtained when the length defined by the Weyl scalar curvature is small if compared to the typical length scale of the system. Once the Weyl curvature becomes non-negligible the system goes into a quantum regime. In this approach, as long as one accepts that quantum mechanics is a manifestation of a non-Euclidean geometry, all theoretical issues related to quantum effects receive a pure geometrical meaning. In particular, the identification of the Weyl integrable space's scalar curvature leads to a geometrical interpretation of Heisenberg's uncertainty principle. The uncertainty principle derives from the fact that we are unable to perform a classical measurement to distances smaller than the Weyl curvature length. In other words, the size of a measurement has to be bigger than the Weyl length

$$\Delta L \geq L_W = \frac{1}{\sqrt{R}}. \tag{1.111}$$

The quantum regime is extreme when the Weyl curvature term dominates. Novello's, Salim's and Falciano's interpretation of Heisenberg's uncertainty principle resembles Bohr's complementary principle because of the impossibility of applying the classical definitions of measurements. However, while Bohr's complementary principle is based on the uncontrolled interference of a classical apparatus of measurement, here it is necessary to include the Weyl curvature, and a classical measurement of distance can no longer be performed.

1.2.5 The Entropic Approach to Quantum Potential

In the 2008's articles *Bohmian split of the Schrödinger equation onto two equations describing evolution of real functions* [36] and *Bohmian trajectories in the path integral formalism. Complexified Lagrangian mechanics* [37] V. I. Sbitnev developed the entropic approach of the quantum potential for one-body systems showing that the formalism leads to a complexified Hamilton-Jacobi equation and that the quantum potential which emerges from the quantum entropy is a term that modifies Feynman's path integrals by expanding coordinates and momenta to the imaginary sector.

In the case of a one-body system, an entropic version of Bohm's quantum potential in non-relativistic regime can be developed by defining the logarithmic function

$$S_Q = -\frac{1}{2}\ln\rho \tag{1.112}$$

where $\rho(\vec{x},t) = R^2(\vec{x},t) = |\psi(\vec{x},t)|^2$ is the probability density (describing the density of particles in the element of volume d^3x around a point $\vec{x}$ at time t) associated with the wave function $\psi(\vec{x},t)$ of the individual physical system. In analogous way, an entropic version of Bohm's quantum potential for a many-body system constituted by N particles starts by a logarithmic function of the form (1.112) where $\rho = |\psi(\vec{x}_1,\vec{x}_2,\ldots,\vec{x}_N,t)|^2$ is the probability density associated with the wave function $\psi(\vec{x}_1, \vec{x}_2, \ldots, \vec{x}_N, t)$ of the many-body system. Equation (1.112) presents some analogy with the standard definition of entropy given by Boltzmann law: it provides indeed the quantum counterpart of a Boltzmann-type law. Since it shows a relation with the wave function, the quantity given by Eq. (1.112) can be appropriately defined as "quantum entropy".

In the paper [36] Sbitnev showed that by introducing the quantity (1.159), the quantum potential for a one-body system (1.4) can be expressed in the following convenient way

$$Q = -\frac{\hbar^2}{2m}(\nabla S_Q)^2 + \frac{\hbar^2}{2m}(\nabla^2 S_Q). \qquad (1.113)$$

In this way, Bohm's quantum Hamilton-Jacobi equation (1.3) regarding the motion of the corpuscle associated with the wave function $\psi(\vec{x}, t)$ becomes:

$$\frac{|\nabla S|^2}{2m} - \frac{\hbar^2}{2m}(\nabla S_Q)^2 + V + \frac{\hbar^2}{2m}(\nabla^2 S_Q) = -\frac{\partial S}{\partial t}. \qquad (1.114)$$

Equation (1.114) provides an energy conservation law in which the term $-\frac{\hbar^2}{2m}(\nabla S_Q)^2$ can be interpreted as the quantum corrector of the kinetic energy $\frac{|\nabla S|^2}{2m}$ of the particle while the term $\frac{\hbar^2}{2m}(\nabla^2 S_Q)$ can be interpreted as the quantum corrector of the potential energy V. On the ground of Sbitnev's results, the quantum potential derives from the quantum entropy describing the degree of order and chaos of the configuration space produced by the density of the ensemble of particles associated with the wave function. On the basis of (1.114), the quantum entropy determines two quantum correctors in the energy of the physical system (of the kinetic energy and of the potential energy respectively) and without these two quantum correctors (linked just to the quantum entropy) the total energy of the system would not be conserved.

Moreover, always following the treatment of Sbitnev in the Ref. [36], by substituting the quantum entropy given by Eq. (1.159) in the continuity Eq. (1.6) one obtains the entropy balance equation

$$\frac{\partial S_Q}{\partial t} = -(\vec{v} \cdot \nabla S_Q) + \frac{1}{2} \nabla \cdot \vec{v} \qquad (1.115)$$

where $\vec{v} = \frac{\nabla S}{m}$ is the particle's speed. In Eq. (1.115), the second term at the right hand describes the rate of the entropy flow due to spatial divergence of the speed. This second term is nonzero in regions where the particle changes direction of movement. Taking into account Brillouin's results about the link between a negative value of S_Q and information [38], Eq. (1.115) can be interpreted as a law which describes the balance of the information flows. Therefore, on the basis of Eqs. (1.113)–(1.115), the quantum potential can be seen as an information channel into the behaviour of the particles.

By introducing the quantum entropy given by Eq. (1.112), it is just this quantity describing the degree of order and chaos of the vacuum—a storage of virtual trajectories supplying optimal ones for particle movement—supporting the density ρ (of the particles associated with the wave function) which produces an active information in the behaviour of the particles. The non-local action of the quantum potential can be itself seen as a consequence of the quantum entropy in virtue of the presence of the Laplace operator of the quantum entropy.

On the other hand, the interpretation of the quantum potential based on the quantum entropy has another important merit: while in the usual interpretation of bohmian mechanics the equations of motion are nonlinear in nature, because of the dependence of the quantum potential on the wave function (different initial conditions yield in fact a different quantum potential), instead here, in the entropic approach, the equations of motion become, in a certain sense, "linear". In fact, here one assumes preliminarily that the density of particles ρ associated with the wave function of the physical system under consideration determines a modification of the geometry of the configuration space defined by the logarithmic function (1.112) and then can say that in this "non-linear" background the equations of motion of the system, given now by Eqs. (1.114 and (1.115) are "linear". As it was shown already by E. R. Bittner, the introduction of the quantum entropy (1.112) allows us to transform a non-linear model into a linear one [39].

Finally, as Sbitnev clearly demonstrated in the paper [37], a relevant advantage of the entropic interpretation of the non-relativistic de Broglie-Bohm theory is to lead to a complexified state space as a fundamental background of quantum processes. By means of an opportune unification of the quantum Hamilton-Jacobi equation (1.114) and the entropy balance Eq. (1.115) a complexified Hamilton-Jacobi equation containing complex kinetic and potential terms can be obtained. And in this new approach the two quantum correction terms of kinetic energy and potential energy both depending on the quantum entropy emerge as the fundamental terms that modify the classical Feynman's path integral by expanding coordinates and momenta to imaginary sector.

In this regard, Sbitnev's treatment in the paper [37] for one-body systems and Fiscaletti's treatment in the papers [40, 41] for many-body systems allow us to obtain the following results. The complexified space is defined by a complexified momentum $(\vec{p}' = \nabla J = \nabla S + i\hbar\nabla S_Q$ for one body-systems and $\vec{p}' = \sum\limits_{i=1}^{N} \nabla_i J =$

$\sum\limits_{i=1}^{N} \nabla_i S + i\hbar\nabla_i S_Q$ for many-body systems, where $J = S + i\hbar S_Q$ is a complexified action) and the complexified coordinates $\vec{x}' = \vec{x} + i\vec{\varepsilon}$ (where $\vec{\varepsilon} = \frac{\hbar}{2m} s\vec{n}$ is a small vector having the dimension of a length, s being the universal constant given by $s = 4\pi\varepsilon_0 \frac{\hbar}{e^2} = 4,57 \cdot 10^{-7} [s/m]$, e being the elementary charge carried by a single electron, ε_0 the vacuum permittivity). In this complexified space the complexified action $J = S + i\hbar S_Q$ is introduced by starting from equations

(a) $\dfrac{|\nabla S|^2}{2m} + i\hbar\dfrac{1}{m}(\nabla S \cdot \nabla S_Q) - \dfrac{\hbar^2}{2m}(\nabla S_Q)^2 + V - \dfrac{1}{2}i\hbar(\nabla\vec{v}) + \dfrac{\hbar^2}{2m}(\nabla^2 S_Q) = -\dfrac{\partial S}{\partial t}$

$$(1.116)$$

(for one-body systems)

$$\text{(b)} \sum_{i=1}^{N} \frac{|\nabla_i S|^2}{2m_i} + i\hbar \sum_{i=1}^{N} \frac{1}{m_i}(\nabla_i S \cdot \nabla_i S_Q) - \sum_{i=1}^{N} \frac{\hbar^2}{2m_i}(\nabla_i S_Q)^2$$

$$+ V - \frac{1}{2} i\hbar \sum_{i=1}^{N} \nabla_i \cdot \vec{v}_i + \sum_{i=1}^{N} \frac{\hbar^2}{2m_i}(\nabla_i^2 S_Q) \tag{1.117}$$

$$= -\frac{\partial S}{\partial t}$$

(for many-body systems)
and satisfies the following complexified Hamilton-Jacobi equations

$$\text{(a)} \ -\frac{\partial J}{\partial t} = \frac{1}{2m}(\nabla J)^2 + V(\vec{x}') = H(\vec{x}', \vec{p}', t) \tag{1.118}$$

and

$$\frac{dJ}{dt} = -H(\vec{x}', \vec{p}', t) + \sum_{i=1}^{N} p_i' \dot{x}_i' = L\left(\vec{x}', \dot{\vec{x}}', t\right) \tag{1.119}$$

for one-body systems;

$$\text{(b)} \ -\frac{\partial J}{\partial t} = \sum_{i=1}^{N} \left[\frac{1}{2m_i}(\nabla_i J)^2 + V(\vec{x}_i')\right] = \sum_{i=1}^{N} H_i\left(\vec{x}_i', \vec{p}_i', t\right) \tag{1.120}$$

and

$$\frac{dJ}{dt} = \sum_{i=1}^{N} L_i\left(\vec{x}', \dot{\vec{x}}', t\right) \tag{1.121}$$

for many-body systems (where $H(\vec{x}', \vec{p}', t)$ is the complexified Hamiltonian, $L\left(\vec{x}', \dot{\vec{x}}', t\right)$ is the complexified Lagrangian),

whose solutions are respectively

$$\text{(a)} \ J = -\int_{t_0}^{t} H(\vec{x}', \vec{p}', \tau)d\tau + C_1 \tag{1.122}$$

$$J = -\int_{t_0}^{t} L\left(\vec{x}', \dot{\vec{x}}', \tau\right)d\tau + C_2 \tag{1.123}$$

where C_1 and C_2 are two integration constants satisfying the following condition:

$$C_1 - C_2 = \int_{t_0}^{t} \sum_{i=1}^{N} p_i' \dot{x}_i' dt = \int_{L} \sum_{i=1}^{N} p_i' dx_i' \tag{1.124}$$

for one-body systems;

$$(b)\; J = -\int_{t_0}^{t} \sum_{i=1}^{N} H_i\left(\vec{x}_i', \vec{p}_i', \tau\right) d\tau + C_1 \tag{1.125}$$

$$J = -\int_{t_0}^{t} \sum_{i=1}^{N} L_i\left(\vec{x}_i', \dot{\vec{x}}_i', \tau\right) d\tau + C_2 \tag{1.126}$$

where C_1 and C_2 are two integration constants satisfying the following condition:

$$C_1 - C_2 = \sum_{i=1}^{N} \int_{L_i} \sum_{k=1}^{N} p_k' dx_k' \tag{1.127}$$

for many-body systems.

The complexified state space deriving from the quantum entropy can be considered as the fundamental stage which determines the features of Bohmian trajectories: Bohmian trajectories are submitted to the principle of least action that expands on the action integral (1.123) for one-body systems (or (1.126) for many-body systems) containing the complexified Lagrangian function derived from the quantum entropy. Bohmian trajectories turn out to be geodesic trajectories of an incompressible fluid loaded by the complexified Lagrangian that in turn is determined by the two quantum correctors fixed by the quantum entropy. Moreover, the two Bohmian correctors modifies the Feynman's path integral by expanding coordinates and momenta to the imaginary sector. In the case of a one-body system, as shown by C. Grosche [42], Feynman's path integral can be written mathematically in the following way inside the complexified state space

$$K\left(\vec{x}', t; \vec{x}_0', t_0\right) = \int \int \cdots \int D[\vec{x}'(\tau)] \exp\left\{ \frac{i}{\hbar} \int_{t_0}^{t} L\left(\vec{x}', \dot{\vec{x}}', \tau\right) d\tau \right\} \tag{1.128}$$

where the path-integral symbol indicates the multiple integral

$$\int \int \cdots \int D[\vec{x}'(\tau)] \Leftrightarrow \left(\frac{2\pi i\hbar \delta t}{m} \right)^{-M/2} \int_{\vec{x}_0'}^{\vec{x}'} d\vec{x}_1' \int_{\vec{x}_0'}^{\vec{x}'} d\vec{x}_2' \cdots \int_{\vec{x}_0'}^{\vec{x}'} d\vec{x}_M'. \tag{1.129}$$

The quantum superposition principle underlies the path integral. Whereas evolution of a classical object is described by a unique trajectory satisfying the principle of least action, the path integral tests all possible virtual classical trajectories, among which there is a unique trajectory satisfying the least action principle. Other trajectories cancel each other by their interference.

On the basis of Sbitnev's results, the interpretation of Feynman's path integral approach based on Eqs. (1.128) and (1.129) seems simple and natural by the complexified momenta $\vec{p}' = \nabla J = \nabla S + i\hbar\nabla S_Q$ and by the complexified coordinates $\vec{x}' = \vec{x} + i\vec{\varepsilon}$: the reading key is provided by the two Bohmian quantum correctors linked to the quantum entropy. The path integral computation stems directly from decomposition of the Schrödinger equation to the modified quantum Hamilton-Jacobi equation plus the entropy balance equation. The two Bohmian quantum correctors linked with the quantum entropy (1.112) and resulted from this decomposition allow the expanding of the state space to the imaginary sector. Thus, one has a non-trivial N-dimensional manifold embedded in the 2N-dimensional complex state space where its real part is the conventional coordinate state space.

On the basis of the mathematical formalism of the complexified space, one may introduce here a minimum principle for the complexified action which allows us to derive non-relativistic quantum mechanics. This minimum principle is Hamilton's principle for the complexified action $\delta J = 0$ which states that the motion of an arbitrary mechanical system occurs in such a way that the integral (1.123) for one-body systems and (1.126) for many-body systems becomes stationary for arbitrary possible variations of the configuration of the system, provided that initial and final configurations of the system are prescribed.

In the entropic interpretation of the quantum potential, Hamilton's principle of least action can be applied to non-relativistic quantum mechanics by considering the wave function in the complexified space as a generalized coordinate and constructing a Lagrangian density such that Hamilton's principle gives the Schrödinger equation. Hamilton's principle of least action can be used to obtain the Schrödinger equation by starting from a general Lagrangian density characterizing the complexified space. By following the treatment provided in [41] for one-body systems, the Lagrangian density is constructed from the wave function of the complexified space

$$\psi(\vec{x}', \vec{p}', t) = \frac{1}{Z_2}\exp\left\{-\frac{i}{\hbar}\int_{t_0}^{t} L\left(\vec{x}', \dot{\vec{x}}', \tau\right)d\tau\right\} \tag{1.130}$$

from its complex conjugate and from their partial derivatives to any order. The Lagrangian density is defined by relation

$$L = \psi^*\left(i\hbar\frac{\partial}{\partial t} - H\right)\psi \tag{1.131}$$

which depends only on ψ^*, ψ and its partial derivatives. The dynamical equations are derived from this Lagrangian density by using Hamilton's principle of least action. Hamilton's principle states that the action functional $J[\psi^*, \psi]$ for any Lagrangian density (1.131) is stationary:

$$J[\psi^*, \psi] = \int dt \int d^3x' L(\psi, \psi^*, \ldots) = stationary \qquad (1.132)$$

where integration is over all time and all complexified space. The action (1.132) is stationary when its variation with respect to ψ^* or ψ (or both) is zero. When specific Lagrangian density (1.131) is substituted in Eq. (1.132) and variation is made with respect to ψ^* one obtains

$$\delta J[\psi^*, \psi] = \int dt \int d^3x' \delta\psi^* \left(i\hbar \frac{\partial}{\partial t} - H \right) \psi = 0. \qquad (1.133)$$

Since the variation $\delta\psi^*$ is arbitrary except for vanishing at the boundaries, Eq. (1.133) leads immediately to the Schrödinger equation

$$H\psi = i\hbar \frac{\partial \psi}{\partial t} \qquad (1.134)$$

where the initial wave function $\psi(0)$ must be specified. If the variation of the action is made with respect to ψ, then integration by parts is needed and the complex conjugate of the Schrödinger equation is obtained.

In the case of time-independent systems, Hamilton's principle reduces to the energy variational principle of time-independent quantum mechanics: the time-independent Schrödinger equation can be obtained from Hamilton's principle of least action (1.132) by using it with a trial wave function. The principle of least action (1.132) can be used with a trial wave function of the form

$$\psi(\vec{x}', t) = \psi(\vec{x}') \exp(-iEt/\hbar - \eta t) \qquad (1.135)$$

where the wave function ψ (having the general form (1.130)) on the right-hand-side is time independent, E is the energy and $\eta > 0$ is a small parameter that can be taken to be zero at the end of the calculation. Using this wave function in the equation for the action (1.132) and doing the time integration, we obtain

$$J[\psi^*, \psi] = \frac{1}{\eta} \int d^3x' \psi^* (E - \hat{H}) \psi = stationary. \qquad (1.136)$$

If this equation is multiplied by $-\eta > 0$ and the constant E is added, we can obtain the energy variational principle for a stationary state

$$\int d^3x' \psi^* \hat{H} \psi - E \left(\int d^3x' \psi^* \psi - 1 \right) = stationary \qquad (1.137)$$

which is independent of η. The energy E is a Lagrangian multiplier that ensures the normalization of the wave function.

Here, if the variation of Eq. (1.137) is made with respect to the function ψ^*, one can obtain the time-independent Schrödinger equation

$$\hat{H}\psi = E\psi \qquad (1.138)$$

In analogous way, the variation with respect to the function ψ leads to the complex conjugate of the time-independent Schrödinger equation (1.138) after integration by parts. The derivation of Schrödinger's equation from Hamilton's principle in the complexified space determined by the quantum entropy can be considered another relevant result of the entropic interpretation of the quantum potential, which allows us to re-read non-relativistic quantum mechanics in an unitary picture too.

1.3 Non-commutative Quantum Geometry and Hiley's Algebra Processes in Pre-space

According to relevant current research, due mainly to the last Bohm and his co-worker Basil Hiley, the quantum geometrodynamics cannot be considered as fundamental but derives from something more primitive. What Hiley considers "more primitive" is an elementary process described by an opportune algebra from which both geometry and material process unfold together. In synthesis, Hiley defines the fundamental process/potentia the "holomovement" and it has two intertwined aspects: the "implicate order" (characterized algebraically) and the "explicate (or manifest) order" which represents all the physics (for example, the space–time geometry) derived from the implicate order. The holomovement is thus the whole ground form of physical phenomena, which contains orders that are both implicate and explicate, where the latter expresses aspects which emerge from the former. From the perspective of the implicate order, rather than point particles being evolved in time under the instantaneous action of the quantum potential or the pilot wave, the fundamental evolution is one of the processes that give origin to explicate structures taking place in space and time. In this picture, in short, particles and pilot waves are not seen as fundamental entities but rather they emerge from an implicate order which is described algebraically.

Since the early sixties David Bohm introduced the notion of a discrete structural process [43–45], in which the basic entities are not matter or fields in interaction in space-time, but a notion of 'structure process' from which the geometry of space-time and its relationship to matter emerge together providing a general framework that could unify the local facets of relativity and the non-local quantum one.

As regards the problem to find a primitive formalism from which both quantum geometry and dynamic processes unfold together, in 1980 [46] Bohm suggested that the new order to understand quantum phenomena would be based on the notion of "process" and called this new order the "implicate order": the quantum potential must be considered an active information source linked to a quantum background, namely just the implicate order. The intention behind the introduction of this new order was simply to develop new physical theories together with the appropriate mathematical formalism that would lead to new insights into the quantum behavior of matter. In this way Bohm in his last years departed from de Broglie's pilot wave: he suggested the necessity to consider non-locality as a more primary fundamental aspects of the physics world then space-time. In last Bohm's view non-locality is a characteristic of the implicate order whose traces we can only observe weakly in the local explicate order of space-time, also particles, which appears "here" or "there" localized during measurement are expression of an "undivided oneness" which can be described by pre-dynamic algebraic structures, similarly to the John Wheeler pre-geometry [47].

The Bohm's ideas on the structure process if the implicate order found today a natural implementation in the noncommutative geometry field [48].

As regards the research line about the implicate order, in the references [49] and [50] Hiley recently suggested that quantum processes evolve not in space-time but in a more general space called pre-space. In this view, the classical space-time would be a sort of statistical approximation and not all the quantum processes can be projected into this space without producing the familiar non-separability "paradoxes", as the non-locality.

In a paper of 1998 B. Hiley and N. Monk show that this could be realized in a very simple algebraic structure, namely the discrete Weyl algebra [51]. The interplay between implicate and explicate order is considered as a process in a purely mathematical sense and it must be taken as fundamental while space-time, fields and matter can be derived from this fundamental process on the basis of the idea that it is describable by elements of an algebra and the relevant structure process is defined by the algebra itself. In particular, Hiley uses the sympletic Clifford algebra which can be constructed on fermion/boson annihilation and creation operators and contains the Heisenberg algebra, a direct connection to the well-known quantum theory.

In synthesis, the basic underlying assumption of Hiley's general approach suggested in 1993 is that the ontology is based on a process that cannot be described explicitly. It can only be described implicitly, hence the terminology "implicate order". In Hiley's view, this implicate order is a structure of relationships and is not some woolly metaphysical construction, it is a precise description of the underlying process, mathematically expressed in terms of a non-commuting algebra which can thus be considered as the fundamental formal structure in order to understand quantum mechanics.

As regards the non-commutative algebra, new relevant developments have recently emerged from the fusion of Hiley's own ideas [43, 52–54] with those of M. de Gosson [55]. It is on these recent developments that we want now to focus our attention. In this regard, we follow Hiley's fundamental work *Non-commutative quantum geometry: a reappraisal of the Bohm approach to quantum theory* [56]. This work summarises the main B. J. Hiley developments that shad a very different light on the formalism first proposed by Bohm (1952). In this article, Hiley showed that quantum mechanics 'lives' in the covering space of symplectomorphisms and thus that the quantum potential emerges from the symplectic space. Contrary to classical mechanics, in quantum mechanics, the algebra of dynamical operators, which carry the symplectic symmetry, is non-commutative. This means, in particular, that we cannot build an x–p phase space out of the eigenvalues of these operators. Indeed, these eigenvalues do not directly satisfy the symplectic symmetry. The symmetry of the eigenvalues is enfolded or implicit in the symplectic symmetry of the operators. All that has relevant consequences for the quantum information: non-local active information cannot be computed according to the Shannon-Turing schema [57]

By inserting the Bohmian Hamiltonian defined as

$$H^\psi = H + Q^\psi \tag{1.139}$$

into the Hamilton-Jacobi equation written in the form

$$\frac{\partial S}{\partial t} + H^\psi(\vec{r}, \nabla_{\vec{r}} S) = 0 \tag{1.140}$$

one sees that a symplectomorphism $f_{t,t_0}^\psi(\vec{r}_0, \vec{p}_0) = (\vec{r}^\psi(t), \vec{p}^\psi(t))$ exists which can be written in the form

$$\frac{d\vec{r}^\psi}{dt} = \nabla_{\vec{p}} H^\psi \tag{1.141}$$

and

$$\frac{d\vec{p}^\psi}{dt} = -\nabla_{\vec{r}} H^\psi = -\nabla_{\vec{r}}(V + Q^\psi). \tag{1.142}$$

Equation (1.141) is simply the guidance condition $\vec{p} = \nabla S$ from which the trajectories are calculated while Eq. (1.142) is the generalisation of Newton's equation of motion which ensures that the momentum is always conserved thanks to the combined action of the classical potential and the quantum potential. Here by writing

$$S(\vec{r}^\psi(t), t) = S_0(x_0) + \int\limits_0^t (\vec{p} \cdot d\vec{r} - H^\psi dt') \tag{1.143}$$

one finds the relation

$$\psi(\vec{r}^\psi(t), t)|d^n \vec{r}^\psi|^{1/2} = \exp\left[\frac{i}{\hbar} S(\vec{r}^\psi(t), t)\right] \psi_0(x_0)|d^n x_0|^{1/2} \tag{1.144}$$

where $\psi(\vec{r}^\psi(t), t)$ is a solution of Schrödinger's equation. On the basis of Eqs. 1.139–1.144, Schrödinger's equation can be regarded as describing the time evolution of the quantum system in the covering space. Moreover, the quantum potential emerges as an entity which defines the mathematical relationship between the phase space and the corresponding covering space. Thus, on the basis of the treatment provided by Eqs. 1.139–1.144, one can see that the quantum potential plays an essential role in the mathematical relationship between the phase space and the corresponding covering space. Moreover, it is the property of covering spaces that ensures that the underlying trajectories do not cross, thus explaining a well-known property of the Bohm trajectories. The picture provided by Eqs. 1.139–1.144 implies that there is nothing ad hoc or artificial in Bohm's theory (and, in particular about the quantum potential). The features of Bohm's approach and of the quantum potential emerge directly from the underlying non-commutative geometry. These fundamental aspects have been explored by Hiley with the description of the covering space of symplectomorphism in terms of a generalisation of the Heisenberg algebra.

The extended Heisenberg algebra used by Hiley in [56] starts by defining the density operator as

$$\hat{\rho} = \hat{\psi}_L \hat{\psi}_R \tag{1.145}$$

where $\hat{\psi}_L = \hat{A}\varepsilon$ is an element of the left ideal (which represents the operator equivalent to the wave function) and $\hat{\psi}_R = \varepsilon\hat{B}$ is an element of the right ideal (which represents the operator equivalent to the complex conjugate wave function). The left ideal $\hat{\psi}_L$ is also called the algebraic sympletic spinor, while the right ideal $\hat{\psi}_R$ is known as the dual sympletic spinor. In Hiley's approach the Heisenberg equation of motion

$$i\frac{d\hat{A}}{dt} + [\hat{H}, \hat{A}] = 0 \tag{1.146}$$

is replaced by the two following operator Schrödinger equations:

$$i\frac{\partial\hat{\psi}_L}{\partial t} = \hat{H}\hat{\psi}_L \tag{1.147}$$

$$-i\frac{\partial\hat{\psi}_R}{\partial t} = \hat{\psi}_R\hat{H}. \tag{1.148}$$

Both the elements of left ideal and of right ideal and the equations of motion (1.147) and (1.148) are independent of the used representation. Now, by taking the difference between Eqs. (1.147) and (1.148) one obtains the Liouville equation (written in terms of the operators):

$$i\frac{\partial\hat{\rho}}{\partial t} + [\hat{\rho}, \hat{H}]_- = 0 \tag{1.149}$$

which can be considered as an equation governing the time evolution of the amplitude of the process. The sum of Eqs. (1.147) and (1.148) gives

$$i\left[\left(\frac{\partial\psi_L}{\partial t}\right)\hat{\psi}_R - \hat{\psi}_L\left(\frac{\partial\hat{\psi}_R}{\partial t}\right)\right] = [\hat{\rho},\hat{H}]_+. \tag{1.150}$$

On the basis of Eqs. (1.149) and (1.150), both the commutator and the anti-commutator are necessary to provide a complete and coherent description of the Schrödinger equation and its dual. Moreover, Eq. (1.150) can be simplified by decomposing in polar form $\hat{\psi}_L$ and $\hat{\psi}_R$ so that one can write:

$$\hat{\psi}_L = \hat{R}\hat{U}; \ \hat{\psi}_R = \hat{U}^+\hat{R} \tag{1.151}$$

where $\hat{R}$ is positive definite and $\hat{U}$ is unitary. Then one finds:

$$i\left[\left(\frac{\partial\psi_L}{\partial t}\right)\hat{\psi}_R - \hat{\psi}_L\left(\frac{\partial\hat{\psi}_R}{\partial t}\right)\right] = i\hat{R}\left[\frac{\partial\hat{U}}{\partial t}\hat{U}^+ - \hat{U}\left(\frac{\partial\hat{U}^+}{\partial t}\right)\right]\hat{R} = [\hat{\rho},\hat{H}]_+ \tag{1.152}$$

which, if one writes $\hat{U} = e^{i\hat{S}}$ where $\hat{S} = \hat{S}^+$ and assumes that $\left[\hat{R},\frac{\partial\hat{S}}{\partial t}\right] = 0$, becomes:

$$\hat{\rho}\left(\frac{\partial\hat{S}}{\partial t}\right) + \frac{1}{2}[\hat{\rho},\hat{H}]_+ = 0. \tag{1.153}$$

Equation (1.153) is a law which describes the time evolution of the phase operator. Therefore, in the extended Heisenberg algebra, the two Schrödinger–type Eqs. (1.147) and (1.148) are replaced by the two Eqs. (1.149) (which governs the time evolution of the amplitude of the process) and (1.153) (which concerns the evolution of the phase operator).

Now, as regards Eqs. (1.149) and (1.153), it is easy to note a similarity with the equations of motion (1.6) and (1.3) of Bohm's original approach. The difference lies in the fact that, while Bohm's equations of motion (1.6) and (1.3) are written in the position representation, Eqs. (1.149) and (1.153) are operator equations and independent of the representation. In this way, the old criticism to the early Bohm's "positional ontology" is wiped out and substituted by the complementary symmetry of phase spaces.

By projecting Eqs. (1.149) and (1.153) into the position representation, these two operator equations directly lead to the two Bohm equations (1.6) and (1.3) and thus Hiley's approach in the extended Heisenberg algebra turns out to be tightly connected and related to the original Bohm work. If one considers a general representation defined by relation

$$\hat{A}|a\rangle = a|a\rangle \tag{1.154}$$

one immediately finds that Eq. (1.149) becomes

$$i\frac{\partial P(a)}{\partial t} - \left\langle [\hat{H},\hat{\rho}]_-\right\rangle_a = 0 \tag{1.155}$$

where $P(a)$ is the probability of finding the particle in the state $|a\rangle$. Hence, by choosing the Hamiltonian to be $H = \frac{p^2}{2m} + V$ and replacing a with x, namely by going to the position representation, Eq. (1.155) (and thus Eq. (1.149)) becomes identical to Bohm's equation (1.6) which expresses the conservation of probability density. In analogous way, in a general representation, Eq. (1.153) becomes

$$P(a)\frac{\partial S(a)}{\partial t} + \frac{1}{2}\left\langle [\hat{H}, \hat{\rho}]_+ \right\rangle_a = 0. \tag{1.156}$$

Equation (1.156) is remarkably similar to the quantum Hamilton-Jacobi equation (1.3) of the primeval Bohm's approach and here the quantum potential is implicitly contained in the anticommutator and emerges directly by choosing a particular Hamiltonian. Here, if one chooses the Hamiltonian for the harmonic oscillator

$$H = \frac{p^2}{2m} + \frac{kx^2}{2}, \tag{1.157}$$

and substitutes it in Eq. (1.156), in the x-representation one obtains

$$\frac{\partial S_x}{\partial t} + \frac{1}{2m}\left(\frac{\partial S_x}{\partial x}\right)^2 + \frac{kx^2}{2} - \frac{1}{2mR_x}\left(\frac{\partial^2 R_x}{\partial x^2}\right) = 0. \tag{1.158}$$

By writing Eq. (1.153) in the x-representation, therefore, since $\frac{\partial S_x}{\partial x} = p$ (expressing Bohm's guidance condition) a quantum potential has been obtained, giving us an expression for the conservation of energy.

If one follows the same procedure in the momentum p-representation, one finds that Eq. (1.153) is equivalent to the following equation

$$\frac{\partial S_p}{\partial t} + \frac{p^2}{2m} + \frac{k}{2}\left(\frac{\partial S_p}{\partial p}\right)^2 - \frac{k}{2R_p}\left(\frac{\partial^2 R_p}{\partial p^2}\right) = 0 \tag{1.159}$$

which is a quantum Hamilton-Jacobi equation in the momentum p-representation, with the appearance of a quantum potential. Equation (1.159)—which derives from Eq. (1.153) regarding the time evolution of the phase operator—shows that in the extended Heisenberg algebra a Bohmian interpretation with a quantum potential can be developed in the momentum p-representation too. Here, equation is just an expression of the conservation of energy in the momentum p-representation. This can easily be shown by taking into consideration the ground state of the harmonic oscillator and observing that both Eqs. (1.158) and (1.159) lead to the well known result $E = \omega/2$. Therefore, on the basis of Hiley's extended algebra, one can conclude that a Bohm's interpretation of non-relativistic quantum mechanics can be constructed for any representation whatsoever.

Moreover, Hiley's extended Heisenberg algebra clearly shows that the quantum potential is not ad hoc but a necessary feature in order to ensure the conservation of both energy and momentum. For example, the kinetic energy which appears in

Eq. (1.158) is calculated from the real part of $\frac{1}{2m}\left[\psi^*(x,t)\hat{P}\psi(x,t)\right]^2$ and the difference $\frac{1}{2m}\psi^*(x,t)\hat{P}^2\psi(x,t) - \frac{1}{2m}\left[\psi^*(x,t)\hat{P}\psi(x,t)\right]^2$ is simply the quantum potential and thus the conservation of energy requires just Eq. (1.158). In analogous way, in the p-representation, the potential energy which derives from the real part of $\left[\phi^*(p,t)\hat{X}\phi(p,t)\right]^2$ cannot be the total potential energy, which must be calculated from $\phi^*(p,t)\hat{X}^2\phi(p,t)$ and the difference between them lies in the presence of the quantum potential $Q_p = -\frac{1}{2R_p}\left(\frac{\partial^2 R_p}{\partial p^2}\right)$ and thus the conservation of energy requires Eq. (1.159).

Other relevant developments in the description of quantum processes obtained by Hiley inside the picture of non-commutative algebras are represented by the Clifford algebra. In the Ref. [58], Hiley introduced a general method to describe all the usual quantum properties of the Schrödinger, Pauli and Dirac particles within the Clifford algebra structure without the need to use Hilbert spaces. In particular, by starting from an element of the minimal left ideal which encodes all the information normally contained in the conventional wave function, Hiley applied the Clifford algebra $C_{0,1}$ to the Schrödinger particle. In this picture, the properties of a quantum process are obtained in terms of specific element of the algebra and the quantum potential emerges directly from the mathematical laws governing appropriate elements of the algebra.

In order to analyse Hiley's contribution as regards the description of Schrödinger particles inside the Clifford algebra structure, we follow the article *The Clifford algebra approach to quantum mechanics A: the Schrödinger and Pauli particles* [59] where the details to describe the Schrödinger particle and the Pauli particle inside the Clifford algebra are presented in a clear and precise way. The starting point of this approach is to introduce an element of a minimal left ideal $\Phi_L(\vec{r},t) = \phi_L(\vec{r},t)\varepsilon$, where ε is a specific primitive idempotent and $\phi_L(\vec{r},t)$ is a linear combination of some of the elements of the algebra which encodes all the information contained in the "traditional" wave function. In order to specify the state of a quantum system one needs to define a quantity given by relation

$$\rho_C = \Phi_L(\vec{r},t)\Phi_R(\vec{r},t) = \phi_L(\vec{r},t)\varepsilon\phi_R(\vec{r},t), \qquad (1.160)$$

where Φ_R—called element of a minimal right ideal—is the conjugate to Φ_L, ϕ_R is the conjugate of ϕ_L. The quantity (1.160) is called the Clifford density element and plays a central role in the approach to quantum mechanics based on the Clifford algebra because it contains all the information necessary to describe the state of a system completely. As regards the treatment of the Schrödinger particle, Hiley took under examination the Clifford algebra $C_{0,1}$ which is generated by the elements $\{1,e\}$ where $e^2 = -1$. Since in $C_{0,1}$ there is only the idempotent $\varepsilon = 1$, one can write

$$\phi_L = R(g_0 + g_1 e) = RU \qquad (1.161)$$

where $g_0(\vec{r}, t)$ and $g_1(\vec{r}, t)$ are scalar real functions, $R(\vec{r}, t)$ is a real scalar. Here the Clifford conjugate is $\Phi_R = R\tilde{U}$ where $\tilde{U} = g_0 - g_1 e$ and thus one obtains that here the Clifford density element becomes

$$\rho_C = \Phi_L(\vec{r}, t)\Phi_R(\vec{r}, t) = \phi_L(\vec{r}, t)\phi_R(\vec{r}, t) = R^2. \qquad (1.162)$$

These results may be related to the conventional approach by using relations $2g_0 = \psi + \psi^*$ and $2eg_1 = \psi - \psi^*$ where ψ is the ordinary wave function. This means that

$$\rho_C = \Phi_L\Phi_R = \phi_R\phi_L = \psi^*\psi = R^2 = \rho. \qquad (1.163)$$

Thus for the Schrödinger particle, the Clifford density element simply corresponds to the probability density.

Now, in the picture of the Clifford background $C_{0,1}$, the properties of the quantum system under consideration are completely specified in terms of peculiar elements of the algebra called respectively bilinear invariants of the first and second kind (which are calculated by using the Clifford density element). In particular, in the Clifford background, the momentum is defined by relation

$$\rho P^j(t) = -\frac{1}{2}i\Phi_L \overset{\leftrightarrow}{\partial}{}^j \Phi_L \qquad (1.164)$$

while the energy is

$$\rho E(t) = \frac{1}{2}i\Phi_L \overset{\leftrightarrow}{\partial}{}^0 \Phi_L \qquad (1.165)$$

where $\overset{\leftrightarrow}{\partial}$ is the operator defining those entities called by T. Takabayasi [60] the bilinear invariants of the second kind:

$$\psi\overset{\leftrightarrow}{\partial}\bar{\psi} = (\partial\psi)\bar{\psi} - \psi(\partial\bar{\psi}). \qquad (1.166)$$

The momentum (1.164) of the Schrödinger particle may be written as

$$2P^j(t) = -e\Omega^j \qquad (1.167)$$

where

$$\Omega^j = 2(\partial^j U)\tilde{U} \qquad (1.168)$$

Here, if one writes $U = \exp(eS)$ so that $g_0 = \cos S$ and $g_1 = \sin S$, one finds $P^j(t) = \partial^j S$ namely $\vec{P}(t) = \nabla S$

which can be identified with the Bohm momentum. In analogous way, the total energy (1.165) in the Clifford background has the form

$$2E(t) = e\left[\Omega_t U\varepsilon\tilde{U} + U\varepsilon\tilde{U}\Omega_t\right] \qquad (1.169)$$

where $\Omega_t = 2(\partial_t U)\tilde{U}$. For the Schrödinger particle, taking account that $\varepsilon = 1$, if one writes $U = \exp(eS)$ Eq. (1.169) becomes

$$E(t) = -\partial_t S \tag{1.170}$$

which is the expression for the energy used in the old Bohm approach. In this way, one obtains that the fundamental quantum properties of a system (momentum and energy) can be completely described from within the algebra without the need to appeal to any Hilbert space representation.

Moreover, in the Clifford background, the dynamical equation corresponding to (1.149) is

$$i\partial_t \rho_c = [H, \rho_c]_- \tag{1.171}$$

which in the Schrödinger case provides the Liouville equation describing the conservation of probability; the dynamical equation corresponding to (1.153) is

$$i\Phi_L \overleftrightarrow{\partial} \Phi_R = [H, \rho_c]_+ \tag{1.172}$$

which, taking account of Eq. (1.165), becomes

$$\rho E(t) = \frac{1}{2}[H, \rho_c]_+ \tag{1.173}$$

which is the quantum Hamilton-Jacobi equation appearing in conventional Bohm's approach to quantum mechanics. Now, by substituting the Hamiltonian $H = \frac{p^2}{2m} + V$ into Eq. (1.173) and using $\vec{P} = \nabla S$, Eq. (1.173) becomes

$$\partial_t S + \frac{(\nabla S)^2}{2m} + Q + V = 0 \tag{1.174}$$

where $Q = -\frac{\hbar^2}{2m}\frac{\nabla^2 R}{R}$ can be immediately recognized as the quantum potential.

In synthesis, in Hiley's approach to non-relativistic quantum mechanics in a Bohmian framework based on the extended Heisenberg algebra and the Clifford algebra, one can conclude that the quantum potential emerges from a more fundamental background—linked with a non-commutative algebra—which underlies the quantum processes. The extended Heisenberg algebra and the Clifford background can be considered as two different and equivalent mathematical ways to express the reality which underlies the features of quantum phenomena, namely as attempts to develop a mathematical formalism for the foreground of the quantum processes and its features depending on the implicate order. The fundamental Hiley's idea is that, in the quantum world, there is not a priori given manifold: the fundamental reality is the algebra and from the algebra the geometry is then abstracted. It is, actually, a subtle mathematical coming back to Bohr complementarity via Bohm. The process described by the two non-commutative algebras is just the implicate order, while the shadow manifolds (such as that corresponding with the conventional Bohm's equations of motion, and thus the traditional

Bohm's quantum potential) are the explicate orders. In the light of Hiley's algebraic approaches, it is the pre-space, the implicate order expressed by the extended Heisenberg background or by the Clifford background

which determines the origin, the features and the action of the quantum potential. For further and exciting developments see [61, 62].

References

1. De Broglie, L.: Solvay Congress (1927), Electrons and photons: rapports et discussions du Cinquime Conseil de Physique tenu Bruxelles du 24 au Octobre 1927 sous les auspices de l'Istitut International de Physique Solvay. Gauthier-Villars, Paris (1928)
2. de Broglie, L.: Une interpretation causale et non linéaire de la mécanique ondulatoire: la théorie de la doble solution. Gauthier-Villars, Paris (1956)
3. de Broglie, L.: The reinterpretation of quantum mechanics. Found. Phys. 1, 5 (1970)
4. Pauli, W.: Électrons et photons: Rapports et discussions du cinquieme conseil de physique, pp. 280–282. Gauthier-Villars, Paris (1928)
5. Bohm, D.: A new suggested interpretation of quantum theory in terms of hidden variables. Part I. Phys. Rev. 85, 166–179 (1952)
6. Bohm, D.: A new suggested interpretation of quantum theory in terms of hidden variables. Part II. Phys. Rev. 85, 180–193 (1952)
7. Bohm, D.: Proof that probability density approaches $|\Psi|^2$ in causal interpretation of the quantum theory. Phys. Rev. 89, 458–466 (1953)
8. Holland, P.R.: The Quantum Theory of Motion. Cambridge University Press, Cambridge (1993)
9. Bohm, D., Hiley, B., Kaloyerou, P.N.: An ontological basis for quantum theory. Phys. Rep. 144, 321–375 (1987)
10. Bohm, D., Hiley, B.: The Undivided Universe. Routledge, London (1993)
11. Philippidis, C., Dewdney, C., Hiley, B.: Quantum interference and the quantum potential. Nuovo Cimento B 52(1), 15–28 (1979)
12. Dewdney, C., Hiley, B.: A quantum potential description of one-dimensional time-dependent scattering from square barriers and square wells. Found. Phys. 12, 27–48 (1982)
13. Honig, W.M., Kraft, D.W., Panarella, E. (eds.): Quantum Uncertainties: Recent and Future Experiments and Interpretations. Nato ASI Series, Plenum Press, New York (1987)
14. Hiley, B.: Some remarks on the evolution of Bohm' proposals for an alternative to standard quantum mechanics. http://www.bbk.ac.uk/tpru/RecentPublications.html (2010)
15. Dürr, D., Goldstein, S., Zanghi, N.: Quantum equilibrium and the origin of absolute uncertainty. J. Stat. Phys. 67, 843–907 (1992)
16. Bohm, D.: Quantum Theory. Routledge, London (1951)
17. Goldstein, S., Berndl, S., Daumer, M., Dürr, D., Zanghì, N.: A survey of Bohmian mechanics. Il Nuovo Cimento 110B, 737–750 (1995)
18. Goldstein, S., Dürr, D., Zanghì, N.: Bohmian mechanics and quantum equilibrium. In: Albeverio, S., Cattaneo, U., Merlini, D. (eds.) Stochastic Processes, Physics and Geometry II, pp. 221–232. World Scientific, Singapore (1995)
19. Goldstein, S.: Bohmian mechanics and the quantum revolution. Synthese 107, 145–165 (1996)
20. Goldstein, S., Dürr, D., Zanghì, N.: Bohmian mechanics as the foundation of quantum mechanics. In: Cushing, J.T., Fine, A., Goldstein, S. (ed.) Bohmian Mechanics and Quantum Theory: An Appraisal. Boston Studies in the Philosophy of Science, vol. 184, pp. 21–44. Kluwer, Dordrecht (1996)

21. Dürr, D., Goldstein, S., Zanghì, N.: Bohmian mechanics and meaning of the wave function. In: Cohen, R. S., Horne, M., Stachel, J. (eds.) Experimental Metaphysics. Quantum Mechanical Studies for Abner Shimony, vol. 1, pp. 25–38. Kluwer, Dordrecht (1997)
22. Allori, V., Zanghì, N.: What is Bohmian mechanics. Intern. J. Theor. Phys. **43**, 1743–1755 (2004)
23. Goldstein, G., Dürr, D., Tumulka, R., Zanghì, N.: Bohmian mechanics. In: Borchert, D.M. (ed.) The Encyclopedia of Philosophy, 2nd edn. Macmillan Reference, London (2006)
24. Goldstein, S., Dürr, D., Tumulka, R., Zanghì, N.: Bohmian mechanics. In: Weinert, F., Hentschel, K., Greenberger, D. (eds.) Compendium of Quantum Physics. Springer, Berlin (2009)
25. Goldstein, S., Tumulka, R., Zanghì, N.: Bohmian trajectories as the foundation of quantum mechanics. In: Chattaraj P. (ed.) Quantum Trajectories, pp. 1–15. Taylor & Francis, Boca Raton (2010)
26. Goldstein S., Teufel, S.: Quantum spacetime without observers: ontological clarity and the conceptual foundations of quantum gravity. In: Callender, C., Huggett, N. (eds.) Physics Meets Philosophy at the Planck Scale, pp. 275–289. Cambridge University Press, Cambridge (reprinted in Dürr, D., Goldstein, S., Zanghì, N.: Quantum Physics Without Quantum Philosophy. Springer, Berlin (2012)) (2001)
27. Goldstein, S., Zanghì, N.: Reality and the role of the wave function. In: Dürr, D., Goldstein, S., Zanghì, N. (eds.) Quantum Physics Without Quantum Philosophy. Springer, Berlin (2012)
28. Esfeld, M., Lazarovici, D., Hubert, M., Dürr, D.: The ontology of Bohmian mechanics. http://philsci-archive.pitt.edu/9381 (2013)
29. Atiq, M., Karamian, M., Golshani, M.: A quasi-Newtonian approach to Bohmian quantum mechanics. Annales de la Fondation Louis de Broglie **34**(1), 67–81 (2009)
30. Schrödinger, E.: Quantizierung als Eigenwertproblem (Erste Mitteilung) (Quantization as a Problem of Proper Values. Part I). Annalen der Physik., **79**, 361 (reprinted in Collected Papers on Wave Mechanics, American Mathematical Society, 3rd Revised edn. (Nov 12 2003) (1926)
31. Abolhasani M., Golshani, M.: The path integral approach in the frame work of causal interpretation. Annales de la Fondation Louis de Broglie, **28**(1), 1–8 (2003)
32. Grössing, G.: The vacuum fluctuation theorem: exact Schrödinger equation via nonequilibrium thermodynamics. Phys. Lett. A **372**, 4556 (2008)
33. Grössing, G.: On the thermodynamic origin of the quantum potential. Physica A **388**(6), 811–823 (2009)
34. Fiscaletti, D.: The geometrodynamic nature of the quantum potential. Ukrainian J. Phys. **57**(5), 560–572 (2012)
35. Novello, M., Salim, J.M., Falciano, F.T.: On a geometrical description of quantum mechanics. Int. J. Geom. Meth. Mod. Phys. **8**(1), 87–98 (2011)
36. Sbitnev, V. I.: Bohmian split of the Schrödinger equation onto two equations describing evolution of real functions. Kvantovaya Magiya, **5**(1), 1101–1111. http://quantmagic.narod.ru/volumes/VOL512008/p1101.html (2008)
37. Sbitnev, V.I.: Bohmian trajectories and the path integral paradigm. Complexified Lagrangian mechanics. Int. J. Bifurcat. Chaos **19**(7), 2335–2346 (2009)
38. Brillouin, L.: Science and Information Theory, 2 Reprint edn. Dover Publications, New York (17 July 2013) (1962)
39. Bittner, E.R.: Quantum tunneling dynamics using hydrodynamic trajectories. J. Chem. Phys. **112**, 9703 (2000)
40. Fiscaletti, D.: The quantum entropy as an ultimate visiting card of the de Broglie-Bohm theory. Ukrainian J. Phys. **57**(9), 946–963 (2012)
41. Fiscaletti, D.: A geometrodynamic entropic approach to Bohm's quantum potential and the link with Feynman's path integrals formalism. Quantum Matter **2**(2), 122–131 (2013)

42. Grosche, C.: Path integrals, hyperbolic spaces, and Selberg trace formulae. World Scientific, Singapore (1996)
43. Bohm, D.: Space, time and quantum theory understood in terms of a discrete structure process. In: Proceedings of the International Conference on Elementary Particles, Kyoto, pp. 252–287 (1965)
44. Bohm, D.: Quantum theory as an indication of a new order in physics part A: the development of new orders as shown through the history of physics. Found. Phys. 1(4), 359–371 (1971)
45. Bohm, D.: Quantum theory as a new order in physics, part B: implicate and explicate order in physical law. Found. Phys. 3(2), 139–155 (1973)
46. Bohm, D.: Wholeness and the Implicate Order. Routledge, London (1980)
47. Wheeler, John A.: Information, physics, quantum: the search for links. In: Zurek, W. (ed.) Complexity, Entropy, and the Physics of Information. Addison-Wesley, Redwood City (1990)
48. Cartier, C.: A mad day's work: from Grothendieck to Connes and Kontsevich: the evolution of concepts of space and symmetry. Bull. Am. Math. Soc. 38, 389–408 (2001)
49. Hiley, B.J.: Non-commutative geometry, the Bohm interpretation and the mind-matter relationship. AIP Conf. Proc. 573(1), 77 (2001)
50. Hiley, B.J., Fernandes, M.: Process and time. In: Atmanspacher, H., Ruhnau, E. (eds.) Time, Temporality, and Now, pp. 365–382. Springer, Berlin (1997)
51. Hiley, B.J., Monk, N.: A unified algebraic approach to quantum theory. Found. Phys. Lett. 11(4), 371–377 (1998)
52. Brown, M.R., Hiley, B.J.: Schrödinger revisited: an algebraic approach. arXiv: quant-ph/0005026 (2000)
53. Hiley, B. J.: From the Heisenberg picture to Bohm: a new perspective on active information and its relation to Shannon information. In: Khrennikov, A. (ed.) Proceedings of Conference Quantum Theory: Reconsideration of Foundations, pp. 141–162. Växjo University Press, Växjo (2002)
54. Hiley, B.J.: Algebraic quantum mechanics, algebraic spinors and Hilbert space. In: Bowden K.G. (ed.) Boundaries, pp. 149–186. Scientific Aspects of ANPA 24, ANPA, London (2003)
55. De Gosson, M.: The Principles of Newtonian and Quantum Mechanics. Imperial College Press, London (2001)
56. Hiley, B.J.: Non-commutative quantum geometry: a reappraisal of the Bohm approach to quantum theory. In: Elitzur, A., Dolev, S., Kolenda, N. (eds.) Quo Vadis Quantum Mechanics?, pp. 306–324. Springer, Berlin (2005)
57. Licata, I.: Emergence and computation at the edge of classical and quantum systems. In: Licata, I., Sakaji A. (Eds.) Physics of Emergence and Organization. World Scientific, Singapore, (2008)
58. Hiley, B. J.: Process, distinction, groupoids and Clifford algebras: an alternative view of the quantum formalism. In: Coecke B. (ed.) New Structures for Physics. Springer Lecture Notes in Physics, Berlin (2009)
59. Hiley, B.J.: The Clifford algebra approach to quantum mechanics A: the Schrödinger and Pauli particles. arXiv:1011.4031 [math-ph] (2010)
60. Takabayashi, T.: Relativistic hydrodynamics of the Dirac matter. Progress Theor. Phys. Suppl. 4, 2–80 (1957)
61. Hiley, B.J., Callaghan, R.E.: Clifford algebras and the Dirac-Bohm quantum Hamilton-Jacobi equation. Found. Phys. 42, 192–208 (2012)
62. Hiley, B.: On the relationship between the Moyal algebra and the quantum operator algebra of von Neumann. arXiv:1211.2098 [quant-ph] (2012)

Chapter 2
The Quantum Potential in Particle and Field Theory Models

2.1 A Bohmian Way to the Klein-Gordon Relativistic Wave Equation

As known, the Klein-Gordon equation describe the behaviour of relativistic spinless particles:

$$\left(\nabla^2 - \frac{1}{c^2}\frac{\partial^2}{\partial t^2}\right)\psi = \frac{m^2 c^2}{\hbar^2}\psi \tag{2.1}$$

where m is the mass of the particle. It can be derived by starting from the relativistic expression of the energy $E^2 = p^2 c^2 + m^2 c^4$ via relations $E \rightarrow i\hbar\frac{\partial}{\partial t}$ and $p_j \rightarrow -i\hbar\frac{\partial}{\partial x^j}$ or by starting from the action

$$A(\psi) = \frac{1}{2}\int d^4 x\left(\partial^\mu \partial_\mu \psi - \frac{m^2 c^2}{\hbar^2}\psi\right) \tag{2.2}$$

where $x^0 = t$, $x = (\vec{x}, t)$. A bohmian reading of Eq. (2.1) may be provided by making the usual polar decomposition of the wave function,

$$\psi = e^{iS/\hbar} \tag{2.3}$$

where R, S are real Lorentz scalar functions. By substituting (2.3) into (2.1), it splits up into the two following real functions, a quantum Hamilton-Jacobi equation

$$\partial_\mu S \partial^\mu S = m^2 c^2 (1 + Q), \tag{2.4}$$

and a continuity equation

$$\partial_\mu j^\mu = 0 \tag{2.5}$$

I. Licata and D. Fiscaletti, *Quantum potential: Physics,*
Geometry and Algebra, SpringerBriefs in Physics,
DOI: 10.1007/978-3-319-00333-7_2, © The Author(s) 2014

where

$$Q = \frac{\hbar^2}{m^2 c^2} \frac{\left(\nabla^2 - \frac{1}{c^2}\frac{\partial^2}{\partial t^2}\right)|\psi|}{|\psi|} \tag{2.6}$$

is the quantum potential and

$$j^\mu = -(R^2/mc)\partial^\mu S \tag{2.7}$$

is the current associated with the wave function of the particle.

The current satisfying the continuity Eq. (2.5) defines a congruence of world lines of an ensemble of particles connected with the wave $\psi(\vec{x}, t)$. The tangent to a world line is given by the 4-velocity u^μ which is defined in terms of the 4-momentum p^μ via the relation

$$Mu^\mu = p^\mu = -\partial^\mu S \tag{2.8}$$

where

$$M = m(1 + Q)^{1/2} \tag{2.9}$$

is a variable quantum mass. Solving $u^\mu = \dfrac{dx^\mu}{d\tau}$ where τ is the proper time would then yield a trajectory $x^\mu = x^\mu(\tau)$ once the initial position of a particle in the ensemble is specifies. The particle acceleration implied by differentiating (2.4) is given by [1], Ref. [8] in Chap. 1:

$$\frac{du^\mu}{d\tau} = \frac{1}{2}\left(c^2\eta^{\mu\nu} - u^\mu u^\nu\right)\partial_\nu \log(1 + Q). \tag{2.10}$$

Equation (2.9) implies that, since the quantum potential can be a negative number, tachyonic solutions would emerge. For this reason, as underlined by A. Shojai and F. Shojai in the papers [2, 3], the quantum Hamilton-Jacobi equation of the form (2.4) cannot be considered as the correct equation of motion regarding relativistic spinless particles in a bohmian framework. Basing on the starting idea that a correct relativistic equation of motion should not only be Poincarè invariant but also give the correct non-relativistic limit, F. Shojai and A. Shojai have developed an interesting bohmian approach to the Klein-Gordon equation in which the quantum Hamilton-Jacobi equation which derives from the decomposition of the wave function in its polar form $\psi = |\psi| \exp\left(\frac{iS}{\hbar}\right)$ has the following form

$$\partial_\mu S \partial^\mu S = m^2 c^2 \exp Q. \tag{2.11}$$

Equation (2.11) is Poincarè invariant and has the correct non-relativistic limit and here the quantum potential is defined as (2.6), while the continuity equation is

$$\partial_\mu(\rho \partial^\mu S) = 0 \tag{2.12}$$

where ρ is the density of particles in the element of volume d^3x around a point $\vec{x}$ at time t associated with the wave function $\psi(\vec{x}, t)$ of the individual physical system under consideration. The quantum Hamilton-Jacobi equation (2.11) shows that in the relativistic regime the mass of a particle is determined and generated by the quantum potential, according to relation

$$M = m\sqrt{\exp Q}. \tag{2.13}$$

The quantity given by (2.13) can be defined as a variable quantum mass of the relativistic spinless particle, which is Poincarè invariant and has the correct non-relativistic limit.

In F. Shojai's and A. Shojai's theory, the particle trajectory can be derived from the guidance formula and by differentiating Eq. (2.11) leading to a Newton's-type equation of motion:

$$M\frac{d^2x^\mu}{d\tau^2} = \left(c^2\eta^{\mu\nu} - u^\mu u^\nu\right)\partial_\nu M \tag{2.14}$$

where $\eta_{\mu\nu} = diag(1, -1, -1, -1)$ is the signature of the Minkowski space-time metric.

As regards Klein-Gordon's relativistic quantum mechanics in a bohmian picture, interesting results have been developed, besides by F. Shojai and A. Shojai, also by G. Bertoldi, A. Faraggi and M. Matone and H. Nikolic.

As regards the Bertoldi-Faraggi-Matone theory the reader can find details in the references [4–28]. This approach, exclusively on the basis of an equivalence principle—which states that all physical systems can be connected by a coordinate transformation to the free situation with vanishing energy—can lead to a quantum stationary Hamilton-Jacobi equation which is a third order nonlinear differential equation providing a representation of quantum mechanics in terms of trajectories. In the relativistic case, this approach starts by taking into consideration the relativistic classical Hamilton-Jacobi equation of the following form

$$\frac{1}{2m}\sum_1^D \left(\partial_k S^{cl}(q, t)\right)^2 + M_{rel}(q, t) = 0 \tag{2.15}$$

where

$$M_{rel}(q, t) = \frac{1}{2mc^2}\left[m^2c^4 - \left(V(q, t) + \frac{\partial}{\partial t}S^{cl}(q, t)\right)^2\right], \tag{2.16}$$

$V(q, t)$ being the potential. In the time-independent case one has $S^{cl}(q, t) = S_0^{cl}(q) - Et$ and therefore Eq. (2.15) becomes

$$\frac{1}{2m}\sum_1^D \left(\partial_k S_0^{cl}\right)^2 + M_{rel} = 0 \tag{2.17}$$

where

$$M_{rel}(q) = \frac{1}{2mc^2}\left[m^2c^4 - (V(q) - E)^2\right].$$ (2.18)

Hence, by using the equivalence principle one arrives at relativistic quantum Hamilton-Jacobi equations of the form

$$\frac{1}{2m}(\nabla S_0)^2 + M_{rel} - \frac{\hbar^2}{2m}\frac{\nabla^2 R}{R} = 0$$ (2.19)

and

$$\nabla \cdot \left(R^2 \nabla S_0\right) = 0$$ (2.20)

which imply the stationary Klein-Gordon equation

$$-\hbar^2 c^2 \nabla^2 \psi + \left(m^2 c^4 - V^2 + 2EV - E^2\right)\psi = 0$$ (2.21)

where $\psi = Re^{iS_0/\hbar}$.

In the time-dependent case the relativistic classical Hamilton-Jacobi equation is

$$\frac{1}{2m}\eta^{\mu\nu}\partial_\mu S^{cl}\partial_\nu S^{cl} + M'_{rel} = 0$$ (2.22)

where

$$M'_{rel}(q, t) = \frac{1}{2mc^2}\left[m^2 c^4 - V^2(q) - 2cV(q)\partial_0 S^{cl}(q)\right]$$ (2.23)

with $q = (ct, q_1, \ldots, q_D)$. Equation (2.22) has the same structure as (2.17) with the Euclidean metric replaced by the Minkowskian one $\eta_{\mu\nu} = diag(-1, 1, \ldots, 1)$. Here, the equivalence principle is implemented by modifying the classical equation by adding a function to be determined, namely

$$\frac{1}{2m}(\partial S)^2 + M_{rel} + Q = 0$$ (2.24)

(where $(\partial S)^2 \approx \sum (\partial_\mu S)^2$ etc....). Since M'_{rel} depends on S^{cl}, here one needs to make the identification

$$M_{rel}(q, t) = \frac{1}{2mc^2}\left[m^2 c^4 - V^2(q) - 2cV(q)\partial_0 S(q)\right]$$ (2.25)

which differs from M'_{rel} since S now replaces S^{cl}. Implementation of the equivalence principle requires that an arbitrary M^a state satisfies the following relation

$$M^b_{rel}(q^b) = (p^b|p^a)M^a_{rel}(q^a) + (q^a; q^b);$$
$$Q^b(q^b) = (p^b|p^a)Q(q^a) - (q^a; q^b)$$ (2.26)

where $(p^b|p) = \eta^{\mu\nu}p^b_\mu p^b_\nu / \eta^{\mu\nu}p_\mu p_\nu$. Here, considering the identity

$$\alpha^2(\partial S)^2 = \frac{\left(\nabla^2 - \frac{1}{c^2}\frac{\partial^2}{\partial t^2}\right)(R\exp(\alpha S))}{R\exp(\alpha S)} - \frac{\left(\nabla^2 - \frac{1}{c^2}\frac{\partial^2}{\partial t^2}\right)R}{R} - \frac{\alpha\partial\cdot(R^2\partial S)}{R^2} \quad (2.27)$$

if R satisfies the continuity equation $\partial\cdot(R^2\partial S) = 0$ one may set $\alpha = \frac{i}{\hbar}$ to obtain

$$\frac{1}{2m}(\partial S)^2 = -\frac{\hbar^2}{2m}\frac{\left(\nabla^2 - \frac{1}{c^2}\frac{\partial^2}{\partial t^2}\right)(R\exp(iS/\hbar))}{R\exp(iS/\hbar)} + \frac{\hbar^2}{2m}\frac{\left(\nabla^2 - \frac{1}{c^2}\frac{\partial^2}{\partial t^2}\right)R}{R}. \quad (2.28)$$

Hence, Eq. (2.23) becomes

$$M_{rel} = \frac{\hbar^2}{2m}\frac{\left(\nabla^2 - \frac{1}{c^2}\frac{\partial^2}{\partial t^2}\right)(R\exp(iS/\hbar))}{R\exp(iS/\hbar)} \quad (2.29)$$

and thus the corresponding quantum potential for the time-dependent case is

$$Q_{rel} = -\frac{\hbar^2}{2m}\frac{\left(\nabla^2 - \frac{1}{c^2}\frac{\partial^2}{\partial t^2}\right)R}{R}. \quad (2.30)$$

In this way, in the time-dependent case the relativistic quantum Hamilton-Jacobi equation becomes

$$\frac{1}{2m}(\partial S)^2 + M_{rel} - \frac{\hbar^2}{2m}\frac{\left(\nabla^2 - \frac{1}{c^2}\frac{\partial^2}{\partial t^2}\right)R}{R} = 0 \quad (2.31)$$

while the continuity equation is

$$\partial\cdot(R^2\partial S) = 0. \quad (2.32)$$

On the basis of the treatment of Klein-Gordon relativistic quantum mechanics suggested by the Bertoldi-Faraggi-Matone theory, the quantum potential (2.30) can be derived by starting from the equivalence principle, which can be thus considered as a fundamental principle describing the behaviour of relativistic spinless particles. As regards the Bertoldi-Faraggi-Matone theory, according to the authors, it is interesting to make a parallelism with Atiq's, Karamian's and Golshani's quasi-Newtonian approach of the quantum potential. If—as we have seen in Sect. 1.2.2—in Atiq's, Karamian's and Golshani's approach the mathematical form (and consequently the physical features) of the quantum potential in non-relativistic quantum mechanics is obtained, in the context of a quasi-Newtonian picture, by minimizing the total energy of ensemble, without appealing to the Schrödinger equation and the wave function, in analogous way, in the Bertoldi-Faraggi-Matone theory, the mathematical form (and consequently the physical features) of the quantum potential is obtained directly by applying the equivalence principle, without appealing to the Klein-Gordon equation and the wave function. Therefore, just like in Atiq's, Karamian's and Golshani's approach in reference to the Schrödinger equation, in the Bertoldi-Faraggi-Matone theory the quantum potential regarding the behaviour of relativistic spinless particles can be considered, in virtue of the equivalence principle, as the basis of the Klein-Gordon equation rather than a consequence of it. They both show a theoretical centrality of the quantum potential.

Moreover, it is important to underline that, in analogy to F. Shojai's and A. Shojai's model, also the Bertoldi-Faraggi-Matone theory introduces important perspectives as regards the problem of generation of masses in the picture of the quantum potential. In the case of the relativistic classical Hamilton-Jacobi equation, on the basis of the fixed point $M(q^0) = 0$, one obtains $m = 0$. As a consequence, the equivalence principle then implies that all the other masses can be generated by the inhomogeneous term in the transformation properties of the M^0 state, i.e. $\frac{1}{2mc^2} = (q^0; q)$. Furthermore, Eq. (2.26) indicates that masses are expressed in terms of the quantum potential on the basis of the following relations

$$\frac{1}{2mc^2} = (p|p^0)Q^0(q^0) - Q(q). \tag{2.33}$$

In this regard, the quantum potential can also be interpreted as a sort of intrinsic self energy in a similar form to the relativistic self energy (see, for example, the Ref. [27]).

Finally, let us consider H. Nikolic's results. In the papers [29–32] Nikolic has developed a Lorentz-covariant Bohmian interpretation of the Klein-Gordon equation which has two important merits: on one hand, it is characterized by a lack of statistical transparency (namely by the fact that the statistical distribution of particle positions cannot be calculated from the wave function alone without the knowledge of particle trajectories) which here emerges as a virtue of the Bohmian interpretation, in the sense that it opens the perspective of experimentally distinguishing its predictions from the predictions of other possible interpretations; and, on the other hand, above all, the equations for Bohmian particle trajectories, although are non-local, can be naturally written in a Lorentz-covariant form without a preferred Lorentz frame leading thus to a Lorentz-covariant Bohmian version of the Klein-Gordon equation. In Nikolic's approach, the Klein-Gordon equation takes the form

$$\left(\partial_0^2 - \nabla^2 + m^2c^2\right)\varphi = 0 \tag{2.34}$$

where $\eta_{\mu v} = diag(-1, 1, \ldots, 1)$. If $\psi = \varphi^+$ and $\psi^* = \varphi^-$ correspond to positive and negative frequency parts of $\varphi = \varphi^+ + \varphi^-$ the particle current can be expressed as $j_\mu = i\psi * \overset{\leftrightarrow}{\partial}_\mu \psi$ (where $a\overset{\leftrightarrow}{\partial}\mu b = a\partial^\mu b - b\partial^\mu a$) which is a conserved quantity ($\partial^\mu j_\mu = 0$) and the particle number is $N = \int d^3x j_0$. Trajectories have the form $\frac{d\vec{x}}{dt} = \frac{\vec{j}(t, \vec{x})}{j_0(t, \vec{x})}$ namely

$$\frac{dx^\mu}{ds} = \frac{j^\mu}{2m\psi * \psi} \tag{2.35}$$

where s is an affine parameter along the curves in the 4-dimensional Minkowski space-time. Hence, for $t = x_0$ in a natural system of units ($c = \hbar = 1$) one arrives at two real quantum Hamilton-Jacobi equations of the form

$$\partial^\mu (R^2 \partial_\mu S) = 0 \tag{2.36}$$

and

$$\frac{(\partial^\mu S)(\partial_\mu S)}{2m} - \frac{m}{2} + Q = 0 \tag{2.37}$$

where the quantum potential is

$$Q = -\frac{1}{2m}\frac{\partial^\mu \partial_\mu R}{R}. \tag{2.38}$$

From Eqs. (2.35, 2.37) and the identity $\dfrac{d}{ds} = \dfrac{dx^\mu}{ds}\partial_\mu$ one also obtains the equation of motion

$$\frac{d^2 x^\mu}{ds^2} = \partial^\mu Q. \tag{2.39}$$

Nikolic's approach based on Eqs. (2.35–2.39) for the particle trajectories turns out to be manifestly non-local and, at the same time, Lorentz covariant: in fact, the trajectories in space-time do not depend on the choice of the affine parameter s. The trajectories are integral curves of the vector fields j^μ and $-\partial^\mu S$.

2.2 The Quantum Potential in Dirac Relativistic Quantum Mechanics

Several attempts about the Dirac Equation in a Bohmian framework aimed to using the expression for the Dirac current to obtain trajectories for the Dirac electron (for example, Bohm and Hiley (Ref. [10] in Chap. 1, [33]), P. Holland (Ref. [8] in Chap. 1) and S. Gull, A. Lasenby and C. Doran [34]). However in none of these cases a quantum Hamilton-Jacobi equation for the conservation of energy has been derived and consequently no relativistic expression for the quantum potential has been found.

The approach developed by B. Hiley and R. Callaghan in [35] follows the philosophy at the core of Hiley's model of Schrödinger's particles analysed in Sect. 1.3. Just like in the non-relativistic case, also this approach regarding Dirac's relativistic quantum mechanics can be completely described within an appropriate Clifford algebra and without the wave function: the information required to reproduce all quantum effects are carried within the algebra itself. Hiley's and Callaghan's model can obtain exact expressions for the Bohm energy–momentum density, a relativistic quantum Hamilton-Jacobi for the conservation of energy which includes an expression for the quantum potential and a relativistic time development equation for the spin vectors of the particle.

In order to construct a Dirac theory in a bohmian picture, Hiley and Callaghan utilized the Clifford algebra $C_{1,3}$ which is the algebra generated by $\{1, \gamma_\mu\}$ where $\mu = 0, 1, 2, 3$ and $[\gamma_\mu, \gamma_\nu] = 2g_{\mu\nu}$. In this approach the Clifford algebra is assumed as fundamental and a space-time manifold is projected out by using the mapping $\eta : \gamma_\mu \to \widehat{e}_\mu$ where $\widehat{e}_\mu$ is a set of orthonormal unit vectors in a vector space $V_{1,3}$, the Minkowski space-time for an equivalent class of Lorentz observers. Hiley and Callaghan have found an energy conservation equation of the form:

$$(\partial^\mu \partial_\mu \Phi_L)\Phi_R + \Phi_L(\partial^\mu \partial_\mu \Phi_R) + 2m^2 \Phi_L \Phi_R = 0 \qquad (2.40)$$

and the following equation for the time evolution of the spin and its components

$$\Phi_L(\partial^\mu \partial_\mu \Phi_R) + (\partial^\mu \partial_\mu \Phi_L)\Phi_R = 0 \qquad (2.41)$$

where Φ_R and Φ_L are two entities of $C_{1,3}$ respectively called minimal left element and minimal right element and are linked by the Clifford density element

$$\rho_C = \Phi_L \Phi_R = \varphi_L \varepsilon_\gamma \varphi_R, \qquad (2.42)$$

Φ_R is the conjugate to Φ_L,

$$\Phi_L = 2\mathrm{Re}\left[\frac{(a - ib)\varepsilon_\gamma + (c - id)\gamma_{23}\varepsilon_\gamma + (h - in)\gamma_{30}\varepsilon_\gamma + (f + ig)\gamma_{01}\varepsilon_\gamma}{4}\right], \qquad (2.43)$$

where

$$\varepsilon_\gamma = \frac{(1 + \gamma_0)(1 + i\gamma_{12})}{4}, \qquad (2.44)$$

while

$$\varphi_L = a + b\gamma_{12} + c\gamma_{23} + d\gamma_{13} + f\gamma_{01} + g\gamma_{02} + h\gamma_{03} + n\gamma_5 \qquad (2.45)$$

is the even part of Φ_L, φ_R is the even part of Φ_R (φ_R is the conjugate of φ_L) and a, b, c, d, f, g, h, n are eight real functions that can be used to specify the quantum state of the Dirac particle. The minimal left element (2.43) encodes all the information that is normally encoded in the wave function which uses four complex numbers. The Clifford density element (2.42) corresponds with $\bar{\psi}\psi$ of the standard Hilbert approach (where $\bar{\psi}$ is the adjoint wave function, ψ is the usual wave function at four components which are related to (2.43) though relations

$$\psi_1 = a - ib, \quad \psi_2 = -d - ic, \quad \psi_3 = h - in, \quad \psi_4 = f + ig. \qquad (2.46)$$

The Clifford density element (2.42) contains thus all the information necessary to describe the state of a system completely.

Here, we focus our attention on the energy conservation equation. By considering the variables :

$$2\rho P^\mu = [(\partial^\mu \varphi_L)\gamma_{012}\varphi_R - \varphi_L\gamma_{012}(\partial^\mu \varphi_R)] \qquad (2.47)$$

And

$$2\rho W^\mu = -\partial^\mu(\varphi_L\gamma_{012}\varphi_R) \qquad (2.48)$$

which describe the geometry of the Clifford background of the Dirac particle and lead to the following relations

$$-\partial^\mu \varphi_L = [P^\mu - W^\mu]\varphi_L \qquad (2.49)$$

$$\partial^\mu \varphi_R = \gamma_{012}\varphi_R[P^\mu + W^\mu], \qquad (2.50)$$

after some simple algebra manipulations, the energy conservation Eq. (2.40) becomes

$$P^2 + W^2 + \left[J\partial_\mu P^\mu - \partial_\mu P^\mu J\right] + \left[J\partial_\mu W^\mu + \partial_\mu W^\mu J\right] - m^2 = 0. \qquad (2.51)$$

Equation (2.51) can be further simplified by splitting it into its Clifford scalar and pseudoscalar parts. The pseudoscalar part of Eq. (2.51) is simply

$$\left[J\partial_\mu P^\mu - \partial_\mu P^\mu J\right] = 0. \qquad (2.52)$$

This relation puts a constraint on the relation between the spin and the momentum of the particle.

The scalar part of Eq. (2.51) is

$$P^2 + W^2 + \left[J\partial_\mu W^\mu + \partial_\mu W^\mu J\right] - m^2 = 0. \qquad (2.53)$$

In order to compare Eq. (2.53) with the energy equation

$$p_\mu p^\mu - m^2 = 0 \qquad (2.54)$$

corresponding to the standard Dirac equation, one needs to express the momentum in terms of the Bohm energy-momentum vector:

$$2\rho P_B^\mu = tr\left[\gamma^0(\phi_L\overset{\leftrightarrow}{\partial}_\mu\gamma_{012}\phi_R)\right]. \qquad (2.55)$$

This can be achieved by using relations

$$4\rho^2 P^2 = \sum_{i=0}^{3} A_{i\nu}A_i^\nu, \qquad (2.56)$$

$$4\rho^2 P^2 = 4\rho^2 P_B^2 + \sum_{i=1}^{3} A_{i\nu}A_i^{\nu} \qquad (2.57)$$

$$4\rho^2 \Pi^2 = \sum_{i=1}^{3} A_{i\nu}A_i^{\nu}, \qquad (2.58)$$

where ρ is defined as

$$\rho(F_{\mu 4}) = \gamma_\mu \qquad (2.59)$$

with $\mu = 0, 1, 2, 3$

$$\rho(F_5) = i, \qquad (2.60)$$

and the A_i are given by

$$
\begin{aligned}
A_0^{\nu} &= -\left(a\overset{\leftrightarrow}{\partial}^{\nu}b + c\overset{\leftrightarrow}{\partial}^{\nu}d + f\overset{\leftrightarrow}{\partial}^{\nu}g + h\overset{\leftrightarrow}{\partial}^{\nu}n\right) \\
A_1^{\nu} &= -\left(a\overset{\leftrightarrow}{\partial}^{\nu}g + b\overset{\leftrightarrow}{\partial}^{\nu}f + c\overset{\leftrightarrow}{\partial}^{\nu}h + d\overset{\leftrightarrow}{\partial}^{\nu}n\right) \\
A_1^{\nu} &= \left(a\overset{\leftrightarrow}{\partial}^{\nu}f - b\overset{\leftrightarrow}{\partial}^{\nu}g - c\overset{\leftrightarrow}{\partial}^{\nu}n + d\overset{\leftrightarrow}{\partial}^{\nu}h\right) \\
A_1^{\nu} &= \left(a\overset{\leftrightarrow}{\partial}^{\nu}n - b\overset{\leftrightarrow}{\partial}^{\nu}h + c\overset{\leftrightarrow}{\partial}^{\nu}f - d\overset{\leftrightarrow}{\partial}^{\nu}g\right)
\end{aligned}
\qquad (2.61)
$$

and $\overset{\leftrightarrow}{\partial}^{\nu}$ is the Takabayasi bilinear invariants of the second kind operator (Ref. [60] in Chap. 1):

$$\psi\overset{\leftrightarrow}{\partial}\overline{\psi} = (\partial\psi)\overline{\psi} - \psi(\partial\overline{\psi}) \qquad (2.62)$$

and

$$J = \phi_L \gamma^0 \phi_R \qquad (2.63)$$

is the axial current.

In this way, Eq. (2.53) may be written as

$$P_B^2 + \Pi^2 + W^2 + \left[J\partial_\mu W^\mu + \partial_\mu W^\mu J\right] - m^2 = 0. \qquad (2.64)$$

Equation (2.64) is the quantum Hamilton-Jacobi equation which regards Dirac's relativistic quantum mechanics inside Hiley's and Callaghan's model and thus can be considered as the fundamental equation of this approach. From Eq. (2.64) the quantum potential for Dirac's relativistic quantum mechanics can be defined as :

$$Q_D = \Pi^2 + W^2 + \left[J\partial_\mu W^\mu + \partial_\mu W^\mu J\right]. \qquad (2.65)$$

Hiley's and Callaghan's treatment shows that the quantum potential is a physical entity of Dirac's relativistic quantum mechanics which emerges directly from the Clifford algebra $C_{1,3}$ for Dirac's particles. In analogy with Hiley's models of non-relativistic quantum mechanics seen in Sect. 1.3, in this model of Dirac's relativistic quantum mechanics, the Clifford algebra $C_{1,3}$ described by the fundamental Eqs. (2.40 and 2.41) is the foreground, the implicate order of quantum processes and the quantum potential may be obtained directly from the algebra constraints.

2.3 The Quantum Potential in Relativistic Quantum Field Theory

We are going to review here some relevant attempts to construct a Bohm's picture of relativistic quantum field theory. In this regard, we follow the original literature of Bohm, Hiley, J. Bell and P. Kaloyerou (e.g. [33, 36], Pref. Ref [11]). Let us consider Bose-Einstein fields in the Schrödinger representation:

$$\{\varphi(\vec{x})\} = \varphi_1(\vec{x}), \varphi_2(\vec{x}), \varphi_3(\vec{x}), \ldots \tag{2.66}$$

The Schrödinger equation for the wave functional $\Psi(\{\varphi(\vec{x})\}, t)$ (here we assume $\hbar = c = 1$) is

$$H\Psi(\{\varphi(\vec{x})\}, t) = i\frac{\partial}{\partial t}\Psi(\{\varphi(\vec{x})\}, t). \tag{2.67}$$

where

$$H = \sum_k \int d^3x \left[-\frac{1}{2}\frac{\delta^2}{\delta\varphi_k^2(\vec{x})} + \frac{1}{2}|\nabla\varphi_k(\vec{x})|^2 \right] + V(\{\varphi(\vec{x})\}) \tag{2.68}$$

is the Hamitonian in the Schrödinger representation.

By writing the polar representation of the wave functional:

$$\Psi(\{\varphi(\vec{x})\}, t) = R(\{\varphi(\vec{x})\}, t)e^{iS(\{\varphi(\vec{x})\}, t)} \tag{2.69}$$

where R and S are two real wave functionals and by inserting this into the Schrödinger Eq. (2.67), one obtains two coupled partial differential functional equations:

$$\frac{\partial S}{\partial t} + \frac{1}{2}\sum_k \int d^3x \left[-\frac{1}{2}\frac{\delta^2}{\delta\varphi_k^2(\vec{x})} + \frac{1}{2}|\nabla\varphi_k(\vec{x})|^2 \right] + V + Q = 0 \tag{2.70}$$

$$\frac{\partial}{\partial t} R^2 + \sum_k \int d^3x \frac{\delta}{\delta \varphi_k(\vec{x})} J_k(\{\varphi(\vec{x})\}, t) = 0. \qquad (2.71)$$

where

$$Q(\{\varphi(\vec{x})\}, t) = -\frac{1}{2} \sum_k \int d^3x \frac{1}{R} \frac{\delta^2 R(\{\varphi(\vec{x})\}, t)}{\delta \varphi_k^2(\vec{x})} \qquad (2.72)$$

is the quantum potential and

$$J_k(\{\varphi(\vec{x})\}, t) = R^2 \frac{\delta S}{\delta \varphi_k(\vec{x})} \qquad (2.73)$$

is the generalized current density in the field space.

In the Bohmian interpretation of quantum field theory as was proposed in [33, 36], Pref. Ref [11] the quantum fields have a deterministic time evolution given by the classical Eq. (2.70) where the fundamental entity which describes the quantum processes regarding them is the quantum potential for Bose-Einstein fields (2.72). In the deterministic interpretation, all quantum uncertainties are a consequence of the ignorance about the actual initial field configuration ($\{\varphi(\vec{x})\}, t_0$). The main reason for the consistency of this interpretation is the fact that Eq. (2.71) represents the continuity equation, which provides that the statistical distribution $\rho(\{\varphi(\vec{x})\}, t)$ of field configurations $\{\varphi(\vec{x})\}$ is given by the quantum distribution $\rho = R^2$ at any time t, provided that $\rho = R^2$ is given by R^2 at some initial time t_0. The initial distribution is arbitrary in principle, and a brilliant work by A. Valentini on a quantum H-theorem [37] explains why the quantum distribution is the most probable.

Interesting results have been obtained recently by H. Nikolic. In the papers [30, 38–40] Nikolic developed a treatment of bohmian particle trajectories in relativistic bosonic quantum field theory. In [31, 41], he provided a treatment of bohmian particle trajectories in relativistic fermionic quantum field theory. Moreover, in the even more fascinating paper [42], he suggested a Bohmian interpretation for the many-fingered-time Tomonaga-Schwinger equation for quantum field theory, which turns out to be relativistically covariant as a consequence of the fact that it does not require a preferred foliation (in fact, here the quantum state is a functional of an arbitrary timelike hypersurface).

Following the papers [30, 38–40] of Nikolic, in relativistic bosonic quantum field theory the operator $\widehat{\phi}(x)$ corresponding to a real scalar field $\phi(x)$ satisfies, in the Heisenberg picture, the following equation:

$$\left(\partial_0^2 - \nabla^2 + m^2 c^2\right) \widehat{\phi} = J\left(\widehat{\phi}\right) \qquad (2.74)$$

where J is a nonlinear function describing the interaction. In the Schrödinger picture the time evolution is determined via the Schrödinger equation in the form

$$H\left[\phi, -i\frac{\delta}{\delta\phi}\right]\Psi(\phi, t) = i\frac{\partial}{\partial t}\Psi(\phi, t) \qquad (2.75)$$

where Ψ is a functional with respect to $\phi(\vec{x})$ and a function of t. A normalized solution of this can be expanded as $\Psi(\phi, t) = \sum\limits_{-\infty}^{+\infty}\tilde{\Psi}_n(\phi, t)$ where the $\tilde{\Psi}_n$ are unnormalized n-particle wave functionals. In Bohm's interpretation the field $\phi(\vec{x})$ has a causal evolution given by equation

$$\left(\partial_0^2 - \nabla^2 + m^2 c^2\right)\phi(x) = J(\phi(x)) - \left(\frac{\delta Q[\phi, t]}{\delta\phi(\vec{x})}\right)_{\phi(\vec{x})=\phi(x)} \qquad (2.76)$$

where

$$Q = -\frac{1}{2|\Psi|}\int d^3x \frac{\delta^2|\Psi|}{\delta\phi^2(\vec{x})} \qquad (2.77)$$

is the quantum potential. The quantum potential (2.77) can be considered as the ultimate dynamic entity which influences the behaviour of bosonic fields inside in Nikolic's model. In this approach, the n particles associated with the wave functional Ψ—and thus deriving from the action of the quantum potential (2.77)—have causal trajectories determined by a generalization of

equation $\dfrac{d\vec{x}}{dt} = \dfrac{\vec{j}(t, \vec{x})}{j_0(t, \vec{x})}$ as

$$\frac{d\vec{x}_{n,j}}{dt} = \left(\frac{\psi_n^*(x^{(n)})\overleftrightarrow{\nabla}_j\psi_n(x^{(n)})}{\psi_n^*(x^{(n)})\overleftrightarrow{\partial}_{t,j}\psi_n(x^{(n)})}\right)_{t_1=\cdots=t_n=t}. \qquad (2.78)$$

where $\psi_n(\vec{x}^{(n)}, t) = \left\langle 0\left|\hat{\phi}(t, \vec{x}_1)\cdots\hat{\phi}(t, \vec{x}_n)\right|\Psi\right\rangle$ is the n-particle wave function. These n-particles have well defined trajectories even when the probability (in the conventional interpretation of quantum field theory) of the experimental detection is equal to zero. In Nikolic's bohmian interpretation of bosonic quantum field theory one can also introduce a causally evolving "effectivity" parameter $e_n[\phi, t]$ defined as

$$e_n[\phi, t] = \frac{\left|\tilde{\Psi}_n[\phi, t]\right|^2}{\sum\limits_{n'}^{\infty}\left|\tilde{\Psi}_{n'}[\phi, t]\right|^2}. \qquad (2.79)$$

The evolution of the effectivity parameter is determined by the evolution of ϕ given via Eq. (2.76) and by the solution $\Psi(\phi, t) = \sum\limits_{-\infty}^{+\infty}\tilde{\Psi}_n(\phi, t)$ of the Schrödinger equation. This parameter might be interpreted as a probability that there are n particles in the system at time t if the field is equal (but not measured) to $\phi(\vec{x})$ at that time and, in Bohm's interpretation, is assumed to be an actual property of the particles guided by the wave function ψ_n. It is a nonlocal hidden variable attributed to the particles and it is introduced to provide a deterministic description of the creation and destruction of particles. The effectivity leads to a picture in which particle trajectories never begin or end. The creation corresponds with a process in which the effectivity $e_n[\phi, t]$ changes continuously from e = 0 to e = 1, while the destruction corresponds with a similar continuous process from e = 1 to e = 0.

In the picture provided—inside bosonic quantum fiend theory—by the effectivity parameter (2.109), both possibilities corresponding with a process of creation and with a processes of destruction seem somewhat artificial. However, as Nikolic showed in the paper [43], a much more elegant picture of the processes of particle creation and destruction may be obtained inside a Bohmian interpretation of bosonic strings. Superluminal velocities and the corresponding motions backwards in time, if may seem undesirable in the particle picture, on the other hand turn out to assume an appealing role in the Bohmian description of string creation and destruction, thus supporting the viability of particle currents that may generate superluminal motions. In the Bohmian interpretation of strings suggested by Nikolic in [43], the string coordinates $X^\mu(\sigma, \tau)$ have a deterministic evolution ruled by the Hamilton-Jacobi equation

$$\frac{\partial X^\mu(\sigma, \tau)}{\partial \tau} = -\eta^{\mu\nu} \frac{\delta S[X]}{\delta X^\nu(\sigma)} \tag{2.80}$$

where S is the phase of the string wave functional $\Psi[X] = R[X]e^{iS[X]}$, σ and τ are parameters which break the manifest world-sheet covariance of the Bohmian interpretation, and the current which determines the trajectories is

$$J_\mu(X, \sigma) = i\Psi^*[X] \frac{\overleftrightarrow{\delta}}{\delta X^\mu(\sigma)} \Psi[X]. \tag{2.81}$$

The Bohmian equation of motion (2.80) contains solutions associated with the same world-sheet topologies, because $\Psi[X]$ contains amplitudes corresponding to different world-sheet topologies, including those which correspond to string splitting. In particular, some solutions describe deterministic processes of string splitting, which are associated with processes of particle creation and destruction. In this way, unlike the Bohmian interpretation of pointlike particles, the Bohmian interpretation of strings does not require an artificial introduction of stochastic singular points at which the particle trajectories begin and end, or an artificial introduction of effectivities. Thus, the Bohmian interpretation of strings provides a very elegant solution to the problem of particle creation and destruction. Moreover, the string's current (2.81) is locally spacelike and corresponds to local superluminal velocities, which are related to string creation and destruction without leading to singular splitting points, provided that $S[X]$ is sufficiently smooth, leading to a smooth right side of (2.80).

Finally, let us consider the Nikolic bohmian covariant interpretation for the many-fingered-time Tomonaga-Schwinger Equation [42]. Let $x = (x^0, \vec{x})$ be spacetime coordinates. Let $\phi(\vec{x})$ be a dynamical field on a timelike Cauchy hypersurface Σ defined via $x^0 = T(\vec{x})$ where $\vec{x}$ denote coordinates on Σ. If $\hat{H}(\vec{x})$ is the Hamiltonian density operator, the dynamics of the field ϕ is described by the Tomonaga-Schwinger equation

$$\hat{H}\Psi[\phi, t] = i\frac{\delta\Psi[\phi, t]}{\delta T(\vec{x})} \tag{2.82}$$

where $\delta T(\vec{x})$ indicates an infinitesimal change of the hypersurface Σ. The quantity $\rho[\phi, T] = |\Psi[\phi, T]|^2$ represents the probability density for the field to have a value ϕ on Σ or equivalently the probability density for the field to have a value ϕ at time T. For simplicity, take a scalar field with

$$\hat{H}(\vec{x}) = -\frac{1}{2}\frac{\delta^2}{\delta\phi^2(\vec{x})} + \frac{1}{2}\left[(\nabla\phi(\vec{x}))^2 + m^2\phi^2(\vec{x})\right]. \tag{2.83}$$

Writing the wave functional in polar form $\Psi = R\exp(iS)$ with R and S real functionals, Eq. (2.82) is equivalent to the two following real equations

$$\frac{1}{2}\left(\frac{\delta S}{\delta\phi(\vec{x})}\right)^2 + \frac{1}{2}\left[(\nabla\phi(\vec{x}))^2 + m^2\phi^2(\vec{x})\right] + Q[\vec{x}, \phi, T] + \frac{\delta S}{\delta T(\vec{x})} = 0 \tag{2.84}$$

$$\frac{\delta\rho}{\delta T(\vec{x})} + \frac{\delta}{\delta\phi(\vec{x})}\left(\rho\frac{\delta S}{\delta T(\vec{x})}\right) = 0 \tag{2.85}$$

where

$$Q[\vec{x}, \phi, T] = -\frac{1}{2R}\frac{\delta^2 R}{\delta\phi^2(\vec{x})} \tag{2.86}$$

is the quantum potential. Here, the bohmian interpretation consists in introducing a deterministic time dependent hidden variable such that the time evolution of this variable is consistent with the probabilistic interpretation of ρ. This can be naturally achieved by introducing a many-fingered-time field $\Phi[\vec{x}, T]$ that satisfies the following many-fingered-time bohmian equations of motion:

$$\frac{\partial\Phi[\vec{x}, t]}{\partial T(\vec{x})} = \left(\frac{\delta S}{\delta\phi(\vec{x})}\right)_{\phi=\Phi}. \tag{2.87}$$

From (2.87) and the quantum many-fingered-time Hamilton-Jacobi equation (2.84), one obtains:

$$\left(\left(\frac{\partial}{\partial T(\vec{x})}\right)^2 - \nabla_x^2 + m^2\right)\Phi[\vec{x}, T] = \left(-\frac{\partial Q[\vec{x}, \phi, T]}{\partial\phi(\vec{x})}\right)_{\phi=\Phi}. \tag{2.88}$$

This can be interpreted as a many-fingered-time Klein-Gordon equation modified with a nonlocal quantum term on the right side. Now, in order to obtain a manifestly covariant theory one introduces parameters (s^1, s^2, s^3) to serve as coordinates on the 3-dimensional manifold Σ in space-time with $X^\mu(\vec{s})$ the embedding coordinates. The induced metric on Σ is

$$q_{ij}(\vec{s}) = g_{\mu\nu}(X(\vec{s})) \frac{\partial X^\mu(\vec{s})}{\partial s^i} \frac{\partial X^\nu(\vec{s})}{\partial s^j}. \tag{2.89}$$

By making the substitutions

$$\vec{x} \to \vec{s}; \quad \frac{\delta}{\delta T(\vec{x})} \to \frac{g^{\mu\nu} \tilde{n}_\nu}{\sqrt{|g^{\alpha\beta} \tilde{n}_\alpha \tilde{n}_\beta|}} \frac{\delta}{\delta X^\mu(\vec{s})} \tag{2.90}$$

where $\tilde{n}(\vec{s}) = \varepsilon_{\mu\alpha\beta\gamma} \dfrac{\partial X^\alpha}{\partial s^1} \dfrac{\partial X^\beta}{\partial s^2} \dfrac{\partial X^\gamma}{\partial s^3}$ is a normal to the surface Σ, the Tomonaga-Schwinger equation (2.82) may be written as

$$\hat{H}(\vec{s}) \Psi[\phi, X] = i n^\mu(\vec{s}) \frac{\delta \Psi[\phi, X]}{\delta X^\mu(\vec{s})} \tag{2.91}$$

and the bohmian equations of motion (2.87) become

$$\frac{\partial \Phi[\vec{s}, T]}{\partial \tau(\vec{s})} = \left(\frac{1}{|q(\vec{s})|^{1/2}} \frac{\delta S}{\delta \phi(\vec{s})} \right)_{\phi=\Phi} \tag{2.92}$$

where

$$\frac{\partial}{\partial \tau(\vec{s})} = \lim_{\sigma_x \to 0} \int_{\sigma_x} d^3(\vec{s}) n^\mu(\vec{s}) \frac{\delta}{\delta X^\mu(\vec{s})}. \tag{2.93}$$

In the same spirit the quantum many-fingered-time Klein-Gordon equation (2.88) may be rewritten as

$$\left(\left(\frac{\partial}{\partial \tau(\vec{s})} \right)^2 - \nabla^i \nabla_i + m^2 \right) \Phi[\vec{s}, X] = \left(-\frac{1}{|q(\vec{s})|^{1/2}} \frac{\partial Q[\vec{s}, \phi, X]}{\partial \phi(\vec{s})} \right)_{\phi=\Phi} \tag{2.94}$$

where ∇_i is the covariant derivative with respect to s^i and

$$Q[\vec{s}, \phi, X] = -\frac{1}{|q(\vec{s})|^{1/2}} \frac{1}{2R} \frac{\delta^2 R}{\delta \phi^2(\vec{s})} \tag{2.95}$$

is the quantum potential. The quantum potential (2.95) describes quantum processes in Nikolic's bohmian covariant interpretation for the many-fingered-time Tomonaga-Schwinger equation for relativistic quantum field theory. It provides a covariant description of relativistic quantum field theory in the sense that it depends on the induced metric (2.89) in the 3-dimensional manifold Σ. Its active information which is responsible for the behaviour of the fields is linked to the induced metric.

We conclude this chapter with some considerations about how one could construct a fully relativistic description of fermionic fields by starting from the ideas of Hiley's and Callaghan's model of Dirac's relativistic quantum mechanics in a Clifford background. Let us consider an operator $\widehat{\phi}(x)$ corresponding with to real scalar field $\phi(x)$ which satisfies, in the Heisenberg picture, the Dirac-type equation:

$$\left(i\gamma^{\mu}\partial_{\mu} - m\right)\widehat{\phi} = I\left(\widehat{\phi}\right) \tag{2.96}$$

where I is a nonlinear function describing the interaction. In the Schrödinger picture the time evolution is determined via the Schrödinger equation in the form

$$H\left[\phi, -i\frac{\delta}{\delta\phi}\right]\Psi(\phi, t) = i\frac{\partial}{\partial t}\Psi(\phi, t) \tag{2.97}$$

where Ψ is a functional with respect to $\phi(\vec{x})$ and a function of t. A normalized solution of this can be expanded as $\Psi(\phi, t) = \sum\limits_{-\infty}^{+\infty}\tilde{\Psi}_{n}(\phi, t)$ where the $\tilde{\Psi}_{n}$ are unnormalized n-particle wave functionals.

In a Clifford background Eq. (2.96) leads to an energy conservation equation of the form:

$$(\partial^{\mu}\partial_{\mu}\Phi_{L})\Phi_{R} + \Phi_{L}(\partial^{\mu}\partial_{\mu}\Phi_{R}) + 2m^{2}\Phi_{L}\Phi_{R} = I \tag{2.98}$$

where I is the element of Clifford algebra corresponding with to non-linear function I. Equation (2.98) can be considered as the generalization of Eq. (2.40) to relativistic fermionic field theory. Like in Hiley's and Callaghan's approach to Dirac's relativistic quantum mechanics, in Eq. (2.98) Φ_{R} and Φ_{L} are two entities of $C_{1,3}$ linked by the Clifford density element

$$\rho_{C} = \Phi_{L}\Phi_{R} = \phi_{L}\varepsilon_{\gamma}\phi_{R}, \tag{2.99}$$

that here regards the wave functional $\Psi[\phi(x), t]$ and corresponds to $|\Psi|^{2}$ of the standard Hilbert approach.

By utilizing Eqs. (2.47–2.50), after some algebra, the energy conservation Eq. (2.98) becomes:

$$P^{2} + W^{2} + \left[J\partial_{\mu}P^{\mu} + \partial_{\mu}P^{\mu}J\right] + \left[J\partial_{\mu}W^{\mu} + \partial_{\mu}W^{\mu}J\right] - m^{2} = I. \tag{2.100}$$

The scalar part of Eq. (2.100) leads to equation

$$P^{2} + W^{2} + \left[J\partial_{\mu}W^{\mu} + \partial_{\mu}W^{\mu}J\right] - m^{2} = I \tag{2.101}$$

which can be written as

$$P_{B}^{2} + \Pi^{2} + W^{2} + \left[J\partial_{\mu}W^{\mu} + \partial_{\mu}W^{\mu}J\right] - m^{2} = I \tag{2.102}$$

where

$$2\rho P_B^\mu = tr\left[\gamma^0(\varphi_L \overleftrightarrow{\partial}_\mu \gamma_{012} \varphi_R)\right] \qquad (2.103)$$

is the Bohm energy–momentum vector,

$$4\rho^2 P^2 = \sum_{i=0}^{3} A_{i\nu} A_i^\nu, \qquad (2.104)$$

$$4\rho^2 \Pi^2 = \sum_{i=1}^{3} A_{i\nu} A_i^\nu, \qquad (2.105)$$

where ρ, A_i $\overleftrightarrow{\partial}^\nu$ and J are defined as usual.

Equation (2.102) is the quantum Hamilton-Jacobi equation for fermionic field theory in a Clifford background and can be considered as the fundamental equation of this approach. From Eq. (2.102) the quantum potential for fermionic relativistic field theory in a Clifford background can be defined as the quantity

$$Q_D = \Pi^2 + W^2 + \left[J\partial_\mu W^\mu + \partial_\mu W^\mu J\right] - I. \qquad (2.106)$$

References

1. Holland, P.R.: Geometry of dislocated de Broglie waves. Found. Phys. **17**(4), 345–363 (1987)
2. Shojai, A., Shojai, F.: About some problems raised by the relativistic form of de Broglie-Bohm theory of pilot wave. Phys. Scr. **64**(5), 413–416 (2001)
3. Shojai, F., Shojai, A.: Understanding quantum theory in terms of geometry. arXiv:gr-qc/0404102 v1 (2004)
4. Bertoldi, G., Faraggi, A., Matone, M.: Equivalence principle, higher dimensional Moebius group and the hidden antisymmetric tensor of quantum mechanics. Class. Quantum Grav. **17**, 3965 (2000)
5. Brown, H., Sjöqvist, E., Bacciagaluppi, G.: Remarks on identical particles in de Broglie-Bohm theory. Phys. Lett. A **251**, 229–235 (1999)
6. Brown, H.: The quantum potential: the breakdown of classical sympletic symmetry and the energy of localisation and dispersion. arXiv:quant-ph/9703007 (1997)
7. Brown, H., Holland, P.: Simple applications of Noether's first theorem in quantum mechanics and electromagnetism. Am. J. Phys. **72**(1), 34 (2004)
8. Dewdney, C., Horton, G.: Relativistically invariant extension of the de Broglie-Bohm theory of quantum mechanics. J. Phys. A: Math. Gen. **35**(47), 10117 (2002)
9. Holland, P.R.: New trajectory interpretation of quantum mechanics. Found. Phys. **38**, 881–911 (1998)
10. Holland, P.R.: Hamiltonian theory of wave and particle in quantum mechanics I: Liouville's theorem and the interpretation of the de Broglie-Bohm theory. Nuovo Cimento B **116**, 1043–1070; Hamiltonian theory of wave and particle in quantum mechanics II: Hamilton-Jacobi theory and particle back-reaction. Nuovo Cimento B **116**, 1143–1172 (2001)
11. Holland, P.R.: Causal interpretation of Fermi fields. Phys. Lett. A **128**, 9–18 (1988)

12. Holland, P.R.: Uniqueness of conserved currents in quantum mechanics. Ann. Phys. **12**(7/8), 446–462 (2003)
13. Holland, P.R.: The de Broglie-Bohm theory of motion and quantum field theory. Phys. Rep. **224**(3), 95–150 (1993)
14. Holland, P.R.: Implications of Lorentz covariance for the guidance equation in two-slit quantum interference. Phys. Rev. A **67**, 062105 (2003)
15. Holland, P.R.: Computing the wavefunction from trajectories: particle and wave pictures in quantum mechanics and their relation. Ann. Phys. **315**(2), 505–531 (2005)
16. Holland, P.R.: Constructing the electromagnetic field from hydrodynamic trajectories. Proc. R. Soc. A **461**, 3659–3679 (2005)
17. Horton, G., Dewdney, C.: A non-local, Lorentz-invariant, hidden variable interpretation of relativistic quantum mechanics based on particle trajectories. J. Phys. A: Math. Gen. **34**, 9871 (2001)
18. Horton, G., Dewdney, C.: A relativistically covariant version of Bohm's quantum field theory for the scalar field. J. Phys. A: Math. Gen. **37**, 11935 (2004)
19. Horton, G., Dewdney, C., Ne'eman, U.: de Broglie's pilot-wave theory for the Klein–Gordon equation and its space-time pathologies. Found. Phys. **32**(3), 463–476 (2002)
20. Horton, G., Dewdney, C., Nesteruk, A.: Time-like flows of energy-momentum and particle trajectories for the Klein-Gordon equation. J. Phys. A: Math. Gen. **33**, 7337 (2000)
21. Mostafazadeh, A., Zamani, F.: Conserved current densities, localisation probabilities, and a new global gauge symmetry of Klein-Gordon fields. arXiv:quant-ph/0312078 (2003)
22. Mostafazadeh, A.: Quantum mechanics of Klein-Gordon-type fields and quantum cosmology. Ann. Phys. **309**(1), 1–48 (2003)
23. Carroll, R.: Quantum Theory, Deformation, and Integrability. North-Holland, Amsterdam (2000)
24. Carroll, R.: Integrable systems as quantum mechanics. arXiv:quant-ph/0309159 (2003)
25. Carroll, R.: (X, ψ) duality and enhanced dKdV on a riemann surface. Nucl. Phys. B **502**(3), 561–593 (1997)
26. Carroll, R.: On the whitham equations and (X, ψ) duality. Supersymmetry and integrable models. lecture notes in physics vol. 502, pp. 33–56. Springer lecture notes in physics (1998)
27. Faraggi, A., Matone, M.: The equivalence postulate of quantum mechanics. Int. J. Mod. Phys. A **15**, 1869–2017 (2000)
28. Matone, M.: The cocycle of quantum Hamilton-Jacobi equation and the stress tensor of CFT. Brazilian. J. Phys. **35**(02A), 316–327 (2005)
29. Nikolic, H.: Covariant canonical quantization of fields and Bohmian mechanics. Eur. Phys. J. C **42**(3), 365 (2005)
30. Nikolic, H.: Bohmian particle trajectories in relativistic bosonic quantum field theory. Found. Phys. Lett. **17**(4), 363–380 (2004)
31. Nikolic, H.: Bohmian particle trajectories in relativistic fermionic quantum field. Found. Phys. Lett. **18**(2), 123–138 (2005)
32. Nikolic, H.: Quantum determinism from quantum general covariance. Int. J. Mod. Phys. **D15**(2006), 2171–2176 (2006)
33. Bohm, D., Hiley, B.J.: On the relativistic invariance of a quantum theory based on beables. Found. Phys. **21**(2), 243–250 (1991)
34. Gull, S., Lasenby, A., Doran, C.: Electron paths, tunnelling and diffraction in the spacetime algebra. Found. Phys. **23**(10), 1329–1356 (1993)
35. Hiley, B.J., Callaghan, R.E.: The Clifford algebra approach to quantum mechanics B: the Dirac particle and its relation to the Bohm approach. arXiv:1011.4033v1 [math-ph] (2010)
36. Kaloyerou, P.N.: The causal interpretation of the electromagnetic field. Phys. Rep. **244**(6), 287–358 (1994)
37. Valentini, A.: Signal-locality, uncertainty, and the subquantum H-theorem. I. Phys. Lett. A **156**, 5–11 (1991)
38. Nikolic, H.: The general-covariant and gauge-invariant theory of quantum particles in classical backgrounds. Int. J. Mod. Phys. D **12**(3), 407 (2003)

39. Nikolic, H.: A general covariant concept of particles in curved background. Phys. Lett. B
    527(1), 119–124 (2002)
40. Nikolic, H.: Erratum to: "a general covariant concept of particles in curved background".
    Phys. Lett. B **529**(3), 265 (2002)
41. Nikolic, H.: There is no first quantization except in the de Broglie-Bohm interpretation.
    arXiv:quant-ph/0307179 (2003)
42. Nikolic, H.: Covariant many-fingered time Bohmian interpretation of quantum field theory.
    Phys. Lett. **A348**, 166–171 (2006)
43. Nikolic, H.: Boson-fermion unification, superstrings, and Bohmian mechanics. Found. Phys.
    39(10), 1109–1138 (2009)

Chapter 3
The Quantum Potential in Gravity and Cosmology

3.1 Bohm Theory in Curved Space-Time

The fundamental role of Bohm's quantum potential in the treatment of gravity and thus in the general-relativistic domain has been explored by various authors (see, for example, the book of R. Carroll *Fluctuations, information, gravity and the quantum potential* [1]). Here we focus our attention on a very interesting recent research of A. Shojai and F. Shojai. These two authors examined the behaviour of particles at spin 0 in a curved space-time, demonstrating that quantum potential contributes to the curvature that is added to the classic one and reveals deep and unexpected connections between gravity and the quantum phenomena (Ref. [2, 3] in Chap. 2). To develop the discussion about this topic we refer to the articles [2–14].

By the analysis of the quantum effects of matter in the framework of Bohmian mechanics, A. Shojai's and F. Shojai's toy model shows that the motion of a spinless particle with quantum effects is equivalent to its motion in a curved space-time. The quantum effects of matter as well as the gravitational effects of matter have geometrical nature and are highly related: the quantum potential can be interpreted as the conformal degree of freedom of the space-time metric and its presence is equivalent to the curved space-time. The presence of the quantum force is just like having a curved space-time which is conformally flat and the conformal factor is expressed in terms of the quantum potential. All this is expressed by an equation of motion of the form

$$\tilde{g}^{\mu\nu}\tilde{\nabla}_\mu S \tilde{\nabla}_\nu S = m^2 c^2 \tag{3.1}$$

where S is the phase of the wave function ψ, $\tilde{\nabla}_\mu$ is the covariant differentiation with respect to the metric

$$\tilde{g}_{\mu\nu} = \frac{M^2}{m^2} g_{\mu\nu} \tag{3.2}$$

(which is a conformal metric) where the quantum mass is

I. Licata and D. Fiscaletti, *Quantum potential: Physics, Geometry and Algebra*, SpringerBriefs in Physics, DOI: 10.1007/978-3-319-00333-7_3, © The Author(s) 2014

$$M^2 = m^2 \exp Q, \qquad (3.3)$$

where

$$Q = \frac{\hbar^2}{m^2 c^2} \frac{\left(\nabla^2 - \frac{1}{c^2}\frac{\partial^2}{\partial t^2}\right)_g |\psi|}{|\psi|} \qquad (3.4)$$

is the quantum potential (in (3.4), of course, c is the light speed and $\hbar$ is Planck's reduced constant).

Shojai's and F. Shojai's approach to the Broglie-Bohm theory in a curved space-time can be obtained by starting from Bohm's version of Klein-Gordon equation developed in Sect. 2.1: the treatment of a particle moving in a curved background can be done by changing the ordinary differentiating ∂_μ with the covariant derivative ∇_μ and by changing the Lorentz metric with the curved metric $g_{\mu\nu}$. With these replacements in Eqs. (2.11) and (2.12) one obtains the following equations of motion of a spinless particle in a curved background:

$$\nabla_\mu(\rho \nabla^\mu S) = 0 \qquad (3.5)$$

$$g^{\mu\nu} \nabla_\mu S \nabla_\nu S = m^2 c^2 \exp Q \qquad (3.6)$$

where

$$Q = \frac{\hbar^2}{m^2 c^2} \frac{\left(\nabla^2 - \frac{1}{c^2}\frac{\partial^2}{\partial t^2}\right)_g |\psi|}{|\psi|} \qquad (3.4)$$

is the quantum potential. Utilizing then a fruitful observation of de Broglie [15], the quantum Hamilton–Jacobi equation can be written as

$$\frac{m^2}{M^2} g^{\mu\nu} \nabla_\mu S \nabla_\nu S = m^2 c^2. \qquad (3.7)$$

From this relation it can be concluded that the quantum effects are equivalent to the change of the space-time metric from $g_{\mu\nu}$ to

$$\tilde{g}_{\mu\nu} = \frac{M^2}{m^2} g_{\mu\nu} \qquad (3.2)$$

which is a conformal transformation. In this way Eq. (3.8) can be written just as

$$\tilde{g}^{\mu\nu} \tilde{\nabla}_\mu S \tilde{\nabla}_\nu S = m^2 c^2 \qquad (3.1)$$

where $\tilde{\nabla}_\mu$ represents the covariant differentiation with respect to the metric $\tilde{g}_{\mu\nu}$. Moreover, in this new curved space-time the continuity equation will assume the form

$$\tilde{g}^{\mu\nu} \tilde{\nabla}_\mu(\rho \tilde{\nabla}^\mu S) = 0. \qquad (3.8)$$

So, the presence of the quantum potential is equivalent to a curved space-time with its metric being given by Eq. (3.2), providing thus a fundamental geometrization of the quantum aspects of matter. On the basis of this model, one can conclude that there is a dual aspect of the role of geometry in physics. The space-time geometry sometimes looks like what we call gravity and sometimes looks like what we understand as quantum behaviours and the quantum potential can be considered the real intermediate between these two aspects. In other words, one can say that the particles determine the curvature of space-time and at the same time the space-time metric is linked with the quantum potential which influences the behaviour of the particles. Quantum potential itself creates a curvature, and thus a deformation of the geometry of space-time, which may have a large influence on the classical contribution to the curvature of the space-time.

Moreover, the particle trajectory is ruled by Newton's equation of motion:

$$M\frac{d^2x^\mu}{d\tau^2} + M\Gamma^\mu_{\nu\kappa}u^\nu u^\kappa = \left(c^2 g^{\mu\nu} - u^\mu u^\nu\right)\nabla_\nu M \qquad (3.9)$$

which shows that it is determined by the deformation of the geometry of space corresponding to the quantum potential. Equation (3.9) reduces to the standard geodesic equation via the above conformal transformation (3.2).

3.2 Bohmian Theories on Quantum Gravity

The next important step which shows the fundamental active role of the quantum potential in redesigning the geometry of physical space in the presence of gravity is represented by the results regarding the quantum gravity domain. In this regard, first of all, one has to mention that, on the basis of some recent research, Bohm's interpretation of canonical quantum gravity turns out to have several useful aspects and merits (Ref. [6] in Chap. 1), [16–19].

Some of them are:

- It leads to time evolution of the dynamical variables whether the wave function depends on time or not. Therefore in Bohmian quantum gravity we have not the time problem.
- In the Bohmian approach, there is no need to normalize the wave function for a single system.
- The classical limit has a well-defined meaning. When the quantum potential is less than the classical potential and the quantum force is less than the classical force we are in the classical domain.
- There is no need to separate the classical observer and the quantum system in the measurement problem. In a Bohmian picture of the measurement process we have two interacting systems, the system and the observer. After the interaction takes place, the wave function of the system is reduced in the state

corresponding to the outcome of the measurement in a causal way (Ref. [6] in Chap. 1, [20]).

• Finally, and this is the point towards which now it is important to focus our attention, in a generalized geometric picture of Bohm's interpretation one can unify the quantum effects and gravity at the fundamental level of quantum gravity [2, 7, 21–24].

As far as this latest point is concerned, F. Shojai and A. Shojai recently developed a toy model of quantum gravity (providing a scalar-tensor picture of the ideas developed in Sect. 3.1) in which the form of the quantum potential and its relation to the conformal degree of freedom of the space-time metric can be derived using the equations of motion. F. Shojai's and A. Shojai's model demonstrates that it is just the quantum gravity equations of motion which make the quantum potential the entity expressing the geometrical properties which influence the behaviour of the particles and which is related to the space-time metric. Thus, it suggests a sort of unification of the gravitational and quantum aspects of matter at the fundamental level of physical reality represented by quantum gravity. For a discussion about this topic we follow the above mentioned references.

In this Bohmian quantum gravity model a general relativistic system consisting of gravity and classical matter can be determined by the action

$$A_{no-quantum} = \frac{1}{2k} \int d^4x \sqrt{-g} R + \int d^4x \sqrt{-g} \frac{\hbar^2}{m} \left(\frac{\rho}{\hbar^2} \partial_\mu S \partial^\mu S - \frac{m^2}{\hbar^2} \rho \right) \quad (3.10)$$

where $\rho = J^0$ is the ensemble density of the particles, $k = 8\pi G$ and hereafter we chose the units in which $c = 1$. The introduction of the quantum effects are equivalent to the change of the space-time metric from $g_{\mu\nu}$ to $g_{\mu\nu} \rightarrow g^I_{\mu\nu} = \frac{g_{\mu\nu}}{\exp Q}$ which is a conformal transformation. In this regard, one can write the action with quantum effects as:

$$A\left[\overline{g}_{\mu\nu}, \Omega, S, \rho, \lambda\right] = \frac{1}{2k} \int d^4x \sqrt{-\overline{g}} (\overline{R}\Omega^2 - 6\overline{\nabla}_\mu \Omega \overline{\nabla}^\mu \Omega) + \int d^4x \sqrt{-\overline{g}} \left(\frac{\rho}{m} \Omega^2 \overline{\nabla}_\mu S \overline{\nabla}^\mu S - m\rho\Omega^4 \right)$$
$$+ \int d^4x \sqrt{-\overline{g}} \lambda \left(\Omega^2 - \left(1 + \frac{\hbar^2 \left(\nabla^2 - \frac{\partial^2}{\partial t^2} \right) \sqrt{\rho}}{m^2 \sqrt{\rho}} \right) \right)$$

$$(3.11)$$

where $\Omega^2 = \exp Q$ is the conformal factor, a bar over any quantity means that it corresponds to no-quantum regime and λ is a Lagrange multiplier introduced in order to identify the conformal factor with its Bohmian value.

By the variation of the above action with respect to $\overline{g}_{\mu\nu}$, Ω, ρ, S and λ we arrive at the following relations as our equations of motion:

1. The equation of motion for Ω:

$$\overline{R}\Omega + 6\left(\overline{\nabla}^2 - \frac{\overline{\partial}^2}{\partial t^2}\right)\Omega + 2\frac{k}{m}\rho\Omega(\nabla_\mu S\overline{\nabla}^\mu S - 2m^2\Omega^2) + 2k\lambda\Omega = 0 \qquad (3.12)$$

2. The continuity equation for the particles:

$$\nabla_\mu\left(\rho\Omega^2\overline{\nabla}^\mu S\right) = 0 \qquad (3.13)$$

3. The equation of motion for the particles:

$$(\nabla_\mu S\overline{\nabla}^\mu S - m^2\Omega^2)\Omega^2\sqrt{\rho} + \frac{\hbar^2}{2m}\left[\left(\overline{\nabla}^2 - \frac{\overline{\partial}^2}{\partial t^2}\right)\left(\frac{\lambda}{\sqrt{\rho}}\right) - \lambda\frac{\left(\overline{\nabla}^2 - \frac{\overline{\partial}^2}{\partial t^2}\right)\sqrt{\rho}}{\rho}\right] = 0$$

$$(3.14)$$

4. The modified Einstein equations for $\overline{g}_{\mu\nu}$:

$$\Omega^2\left[\overline{R}_{\mu\nu} - \frac{1}{2}\overline{g}_{\mu\nu}\overline{R}\right] - \left[\overline{g}_{\mu\nu}\left(\overline{\nabla}^2 - \frac{\overline{\partial}^2}{\partial t^2}\right) - \nabla_\mu\nabla_\nu\right]\Omega^2 - 6\overline{\nabla}_\mu\Omega\overline{\nabla}_\nu\Omega + 3\overline{g}_{\mu\nu}\overline{\nabla}_\alpha\Omega\overline{\nabla}^\alpha\Omega$$

$$+ \frac{2k}{m}\rho\Omega^2\overline{\nabla}_\mu S\overline{\nabla}_\nu S - \frac{k}{m}\rho\Omega^2\overline{g}_{\mu\nu}\overline{\nabla}_\alpha S\overline{\nabla}^\alpha S + km\rho\Omega^4\overline{g}_{\mu\nu}$$

$$+ \frac{k\hbar^2}{m^2}\left[\overline{\nabla}_\mu\sqrt{\rho}\overline{\nabla}_\nu\left(\frac{\lambda}{\sqrt{\rho}}\right) + \overline{\nabla}_\nu\sqrt{\rho}\overline{\nabla}_\mu\left(\frac{\lambda}{\sqrt{\rho}}\right)\right] - \frac{k\hbar^2}{m^2}\overline{g}_{\mu\nu}\overline{\nabla}_\alpha\left[\lambda\frac{\overline{\nabla}^\alpha\sqrt{\rho}}{\sqrt{\rho}}\right] = 0$$

$$(3.15)$$

5. The constraint equation:

$$\Omega^2 = 1 + \frac{\hbar^2}{m^2}\frac{\left(\overline{\nabla}^2 - \frac{\overline{\partial}^2}{\partial t^2}\right)\sqrt{\rho}}{\sqrt{\rho}} \qquad (3.16)$$

The equations of motion (3.12), (3.13), (3.14), (3.15) and (3.16) illustrate the quantum geometrodynamics effects: they tell us that there are back-reaction effects of the quantum factor on the background relativistic geometry. Moreover, in F. Shojai's and A. Shojai's model, by combining Eqs. (3.12) and (3.13) it is possible to arrive at a more simple relation instead of (3.12). If we use the trace of (3.13) and use (3.14), after some mathematical manipulations, we have:

$$\lambda = \frac{\hbar^2}{m^2}\nabla_\mu\left[\lambda\frac{\overline{\nabla}^\mu\sqrt{\rho}}{\sqrt{\rho}}\right]. \qquad (3.17)$$

If one solves this equation in perturbative way in terms of the parameter $\alpha = \frac{\hbar^2}{m^2}$ by writing $\lambda = \lambda^{(0)} + \alpha\lambda^{(1)} + \alpha^2\lambda^{(2)} + \cdots$ and $\sqrt{\rho} = \sqrt{\rho}^{(0)} + \alpha\sqrt{\rho}^{(1)} + \alpha^2\sqrt{\rho}^{(2)} + \cdots$ one will obtain

$$\lambda^{(0)} = \lambda^{(1)} = \lambda^{(2)} = \cdots = 0. \tag{3.18}$$

Thus the perturbative solution of (3.17) is $\lambda = 0$ which is its trivial solution. In this way the equations of quantum gravity become:

$$\overline{\nabla}_\mu\left(\rho\Omega^2\overline{\nabla}^\mu S\right) = 0 \tag{3.19}$$

$$\overline{\nabla}_\mu S\overline{\nabla}^\mu S = m^2\Omega^2 \tag{3.20}$$

$$G_{\mu\nu} = -kT_{\mu\nu}^{(m)} - kT_{\mu\nu}^{(\Omega)} \tag{3.21}$$

where $T_{\mu\nu}^{(m)}$ is the matter energy–momentum tensor and

$$kT_{\mu\nu}^{(\Omega)} = \frac{\left[g_{\mu\nu}\left(\nabla^2 - \frac{\partial^2}{\partial t^2}\right) - \nabla_\mu\nabla_\nu\right]\Omega^2}{\Omega^2} + 6\frac{\nabla_\mu\Omega\nabla_\nu\Omega}{\omega^2} - 3g_{\mu\nu}\frac{\nabla_\alpha\Omega\nabla^\alpha\Omega}{\Omega^2} \tag{3.22}$$

and

$$\Omega^2 = 1 + \alpha\frac{\overline{\left(\nabla^2 - \frac{\partial^2}{\partial t^2}\right)\sqrt{\rho}}}{\sqrt{\rho}}. \tag{3.23}$$

It can be noted that Eq. (3.20) is a Bohmian-type equation of motion, and if we write it in terms of the physical metric $g_{\mu\nu}$, it reads as

$$\nabla_\mu S\nabla^\mu S = m^2 c^2. \tag{3.24}$$

The next step is to make dynamical the conformal factor and the quantum potential. In this regard, by starting from the most general scalar-tensor action

$$A = \int d^4x\left\{\phi R - \frac{\omega}{\phi}\nabla^\mu\phi\nabla_\mu\phi + 2\Lambda\phi + L_m\right\} \tag{3.25}$$

in which ω is a constant independent of the scalar field ϕ, Λ is the cosmological constant, and L_m is the matter Lagrangian (which is assumed to be in the form

$$L_m = \frac{\rho}{m}\phi^a\nabla^\mu S\nabla_\mu S - m\rho\phi^b - \Lambda(1 + Q)^c \tag{3.26}$$

in which a, b, and c are constants), using a perturbative expansion for the scalar field and the matter distribution density as $\phi = \phi_0 + \alpha\phi_1 + \cdots$ and $\sqrt{\rho} = \sqrt{\rho_0} + \alpha\sqrt{\rho_1} + \cdots$ (and imposing opportune physical constraints in order to determine the parameters a, b, and c), F. Shojai and A. Shojai found the following quantum gravity equations:

$$\phi = 1 + Q - \frac{\alpha}{2}\left(\nabla^2 - \frac{\partial^2}{\partial t^2}\right)Q \tag{3.27}$$

$$\nabla^\mu S \nabla_\mu S = m^2\phi - \frac{2\Lambda m}{\rho}(1+Q)(Q-\tilde{Q})$$
$$+ \frac{\alpha\Lambda m}{\rho}\left[\left(\nabla^2 - \frac{\partial^2}{\partial t^2}\right)Q - 2\nabla_\mu Q\frac{\nabla^\mu\sqrt{\rho}}{\sqrt{\rho}}\right] \tag{3.28}$$

$$\nabla_\mu(\rho\nabla^\mu S) = 0 \tag{3.29}$$

$$G^{\mu\nu} - \Lambda g^{\mu\nu} = -\frac{1}{\phi}T^{\mu\nu} - \frac{1}{\phi}\left[\nabla^\mu\nabla^\nu - g^{\mu\nu}\left(\nabla^2 - \frac{\partial^2}{\partial t^2}\right)\right]\phi + \frac{\omega}{\phi^2}\nabla^\mu\phi\nabla^\nu\phi$$
$$-\frac{1}{2}\frac{\omega}{\phi^2}g^{\mu\nu}\nabla^\alpha\phi\nabla_\alpha\phi \tag{3.30}$$

where $\tilde{Q} = \alpha\frac{\nabla_\mu\sqrt{\rho}\nabla^\mu\sqrt{\rho}}{\sqrt{\rho}}$ and $T^{\mu\nu} = -\frac{1}{\sqrt{-g}}\frac{\delta}{\delta g_{\mu\nu}}\int d^4x\sqrt{-g}L_m$ is the energy–momentum tensor.

This geometric quantum gravity model suggested by F. Shojai and A. Shojai (and synthesized in Eqs. (3.27), (3.28), (3.29), (3.30)) allows us to draw some important conclusions:

- According to Eq. (3.30), the causal structure of the space-time metric $g^{\mu\nu}$ is determined by the gravitational effects of matter. On the basis of Eq. (3.27) quantum effects determine directly the scale factor of space-time.
- The mass field given by the right-hand side of Eq. (3.28) consists of two parts. The first part, which is proportional to α, is a purely quantum effect, while the second part, which is proportional to $\alpha\Lambda$, is a mixture of the quantum effects and the large scale structure introduced via the cosmological constant.
- In this model, the scalar field produces the quantum force which appears on the right hand and violates the equivalence principle (just like, in Kaluza-Klein theory, the scalar field-dilaton—produces a fifth force which leads to the violation of the equivalence principle [25]).

Moreover, according to the geodesic Eq. (3.27), the appearance of quantum mass provides a justification of Mach's principle leading to the existence of an interrelation between the global properties of the universe (space-time structure, the large scale structure of the universe) and its local properties (local curvature, motion in a local frame, etc.). The geometry is altered globally (in conformity with Mach's principle) as a consequence of the changes of the quantum potential acting on the geometry determined by a local variation of the matter field distribution. In this sense the Bohmian approach to quantum gravity is highly non–local as it is forced by the nature of the quantum potential. What we call geometry is only the gravitational and quantum effects of matter as a consequence of the presence of the quantum potential which turns out to be the real intermediate between them.

Thus conformally related frames measure different quantum masses and different curvatures. In particular, it is possible to consider two specific frames. One contains the quantum mass field (appearing in the quantum Hamilton–Jacobi equation) and the classical metric while the other contains the classical mass (appearing in the classical Hamilton–Jacobi equation) and the quantum metric. In other frames both the space-time metric and mass field have quantum properties. In virtue of this argument, one can say that different conformal frames are equivalent pictures of the gravitational and quantum phenomena.

Considering the quantum force, the conformally related frames are not distinguishable just like, as regards the treatment of gravity in general relativity, different coordinate systems are equivalent. Since the conformal transformation changes the length scale locally, we measure different quantum forces in different conformal frames. This is analogous to what happens in general relativity in which general coordinate transformation changes the gravitational force at any arbitrary point. Then, the following basic question becomes natural. Does applying the above correspondence, between quantum and gravitational forces, and between the conformal and general coordinate transformations, means that the geometrization of quantum effects implies conformal invariance just as gravitational effects imply general coordinate invariance?

In analogous way to what happens in general relativity, according to F. Shojai's and A. Shojai's approach to quantum gravity in the context of Bohmian theory, at any point (or even globally) the quantum effects of matter can be removed by a suitable conformal transformation. Thus in that point(s) matter behaves classically. In this way one can introduce here a quantum equivalence principle, similar to the standard equivalence principle, that can be called the conformal equivalence principle. According to the quantum equivalence principle gravitational effects can be removed by considering a freely falling frame while quantum effects can be eliminated by choosing an appropriate scale. The latter interconnects gravity and general covariance while the former has the same role about quantum and conformal covariance. On the basis of both these principles, there is no preferred frame, either coordinate or conformal. And these aspects of the geometry of physical space regarding frames characterized by quantum and conformal covariance derive just from the quantum potential.

One can see here that Weyl geometry provides supplementary degrees of freedom which can be identified with quantum effects and seems to create a unified geometric framework for understanding both gravitational and quantum forces. Some features of this picture are the following: (i) Quantum effects appear independent of any length scale. (ii) The quantum mass of a particle is a field. (iii) The gravitational constant is also a field depending on the matter distribution via the quantum potential (cfr. Ref. [3] in Chap. 2, [21, 22]). (iv) A local variation of matter field distribution changes the quantum potential acting on the geometry and alters it globally; the nonlocal character is associated with the quantum potential (cf. [23]).

In order to analyse the link between the Bohmian toy model of F. Shojai and A. Shojai and Weyl geometry, following the Ref. [3] in Chap. 2, we start from the Weyl-Dirac action

$$\Im = \int d^4x \sqrt{-g}\left(F_{\mu\nu}F^{\mu\nu} - \beta^2{}^W R + (\sigma + 6)\beta_{;\mu}\beta^{;\mu} + L_m\right) \tag{3.31}$$

where L_m is the matter Lagrangian, $F_{\mu\nu}$ is the curl of the Weyl 4-vector ϕ_μ, σ is an arbitrary constant and β is a scalar field of weight -1. The symbol ";" represents a covariant derivative under general coordinates and conformal transformations (Weyl covariant derivative) defined as $X_{;\mu} = {}^W\nabla_\mu X - N\phi_\mu X$ where N is the Weyl weight of X. The equations of motion are then

$$\Phi^{\mu\nu} = -\frac{8\pi}{\beta^2}\left(I^{\mu\nu} + N^{\mu\nu}\right) + \frac{2}{\beta}\left(g^{\mu\nu}{}^W\nabla^\alpha{}^W\nabla_\alpha\beta - {}^W\nabla^\mu{}^W\nabla^\nu\beta\right) +$$

$$\frac{1}{\beta^2}\left(4\nabla^\mu\beta\nabla^\nu\beta - g^{\mu\nu}\nabla^\alpha\beta\nabla_\alpha\beta\right) + \frac{\sigma}{\beta^2}\left(\beta^{;\mu}\beta^{;\nu} - \frac{1}{2}g^{\mu\nu}\beta^{;\alpha}\beta_{;\alpha}\right);$$

$${}^W\nabla_\mu F^{\mu\nu} = \frac{1}{2}\sigma\left(\beta^2\phi^\mu + \beta\nabla^\mu\beta\right) + 4\pi J^\mu; \tag{3.32}$$

$$R = -(\sigma + 6)\frac{{}^W\left(\nabla^2 - \frac{1}{c^2}\frac{\partial^2}{\partial t^2}\right)\beta}{\beta} + \sigma\phi_\alpha\phi^\alpha - {}^W\nabla^\alpha\phi_\alpha + \frac{\psi}{2\beta}$$

where

$$N^{\mu\nu} = \frac{1}{4\pi}\left(\frac{1}{4}g^{\mu\nu}F^{\alpha\beta}F_{\alpha\beta} - F^\mu_\alpha F^{\nu\alpha}\right) \tag{3.33}$$

and

$$8\pi I^{\mu\nu} = \frac{1}{\sqrt{-g}}\frac{\delta\sqrt{-g}L_m}{\delta g_{\mu\nu}}; \; 16\pi J^\mu = \frac{\delta L_m}{\delta\phi_\mu}; \psi = \frac{\delta L_m}{\delta\beta} \tag{3.34}$$

For the equations of motion of matter and the trace of the electromagnetic tensor one can use invariance of the action under coordinate and gauge transformations, leading to equations

$${}^W\nabla_\nu I^{\mu\nu} - I\frac{\nabla^\mu\beta}{\beta} = J_\alpha\phi^{\alpha\mu} - \left(\phi^\mu + \frac{\nabla^\mu\beta}{\beta}\right){}^W\nabla_\alpha J^\alpha;$$

$$16\pi I + 16\pi{}^W\nabla_\mu J^\mu - \beta\psi = 0 \tag{3.35}$$

The first relation of (3.35) is a geometrical identity (Bianchi identity) while the second shows the mutual dependence of the field equations. In this Weyl-Dirac approach the gravity fields $g_{\mu\nu}$ and ϕ_μ and the quantum mass field determine the space-time geometry. Here, starting from Eqs. (3.31), (3.32), (3.33), (3.34) and (3.35) one can build a Bohmian quantum gravity which is conformally invariant in the framework of Weyl geometry. If the model has mass this must be a field (since mass has non-zero Weyl weight). The Weyl-Dirac action is a general Weyl invariant action as above and for simplicity now we assume that the matter

Lagrangian does not depend on the Weyl vector so that $J_\mu = 0$. The equations of motion are then

$$\Phi^{\mu\nu} = -\frac{8\pi}{\beta^2}\left(I^{\mu\nu} + N^{\mu\nu}\right) + \frac{2}{\beta}\left(g^{\mu\nu\,W}\nabla^{\alpha\,W}\nabla_\alpha\beta - {}^W\nabla^\mu{}^W\nabla^\nu\beta\right)$$

$$+ \frac{1}{\beta^2}\left(4\nabla^\mu\beta\nabla^\nu\beta - g^{\mu\nu}\nabla^\alpha\beta\nabla_\alpha\beta\right) + \frac{\sigma}{\beta^2}\left(\beta^{;\mu}\beta^{;\nu} - \frac{1}{2}g^{\mu\nu}\beta^{;\alpha}\beta_{;\alpha}\right);$$

$$^W\nabla_\nu F^{\mu\nu} = \frac{1}{2}\sigma\left(\beta^2\phi^\mu + \beta\nabla^\mu\beta\right);$$

$$R = -(\sigma+6)\frac{{}^W\left(\nabla^2 - \frac{1}{c^2}\frac{\partial^2}{\partial t^2}\right)\beta}{\beta} + \sigma\phi_\alpha\phi^\alpha - \sigma\,{}^W\nabla^\alpha\phi_\alpha + \frac{\psi}{2\beta}. \qquad (3.36)$$

The symmetry conditions are

$$^W\nabla_\nu I^{\mu\nu} - I\frac{\nabla^\mu\beta}{\beta} = 0;$$

$$16\pi I - \beta\psi = 0. \qquad (3.37)$$

where $I = I^{\mu\nu}_{\mu\nu}$. Now, if one introduces a quantum mass field, by showing that it is proportional to the Dirac field one can see that this Weyl approach is related to the geometrodynamic features of Bohmian quantum potential. Thus using (3.36) and (3.37) one has

$$\left(\nabla^2 - \frac{1}{c^2}\frac{\partial^2}{\partial t^2}\right)\beta + \frac{1}{6}\beta R = \frac{4\pi}{3}\frac{I}{\beta} + \sigma\beta\phi_\alpha\phi^\alpha + 2(\sigma-6)\phi^\gamma\nabla_\gamma\beta + \frac{\sigma}{\beta}\nabla^\mu\beta\nabla_\mu\beta.$$

$$(3.38)$$

Equation (3.38) can be solved iteratively via

$$\beta^2 = \frac{8\pi I}{R} - \left\{\frac{1}{[(R/6) - \sigma\phi_\alpha\phi^\alpha]}\right\}\beta\left(\nabla^2 - \frac{1}{c^2}\frac{\partial^2}{\partial t^2}\right)\beta + \cdots \qquad (3.39)$$

Now assuming $I^{\mu\nu} = \rho u^\mu u^\nu$ we multiply (3.37) by u^μ and sum to get

$$^W\nabla_\nu(\rho u^\nu) - \rho\left(u_\mu\frac{\nabla^\mu\beta}{\beta}\right) = 0. \qquad (3.40)$$

Then if one inserts (3.37) into (3.40) one obtains

$$u^{\nu\,W}\nabla_\nu u^\mu = \left(\frac{1}{\beta}\right)(g^{\mu\nu} - u^\mu u^\nu)\nabla_\nu\beta. \qquad (3.41)$$

Moreover, from (3.39) one has

$$\beta^{2(1)} = \frac{8\pi I}{R}; \beta^{2(2)} = \frac{8\pi I}{R} \left(1 - \frac{1}{\left(\frac{R}{6}\right) - \sigma\phi_\alpha\phi^\alpha} \frac{\left(\nabla^2 - \frac{1}{c^2}\frac{\partial^2}{\partial t^2}\right)\sqrt{I}}{\sqrt{I}} \right); \cdots . \tag{3.42}$$

Comparing with (3.9) and (3.4) shows that we have the correct equations for the Bohmian theory provided one identifies

$$\beta \approx M; \frac{8\pi I}{R} = m^2; \frac{1}{\sigma\phi_\alpha\phi^\alpha - \frac{R}{6}} \approx \alpha = \frac{\hbar^2}{m^2 c^2} . \tag{3.43}$$

Thus β is the Bohmian quantum mass field and the coupling constant α (which depends on $\hbar$) is also a field, related to geometrical properties of space-time. There is a connection between quantum effects and the length scale of the space-time because the gauge in which the quantum mass is constant (and the quantum force is zero) and the gauge in which the quantum mass is space-time dependent are related each other via a scale change of the form

$$\beta = \beta_0 \rightarrow \beta(x) = \beta_0 \exp(-\Xi(x)); \phi_\mu \rightarrow \phi_\mu + \partial_\mu \Xi. \tag{3.44}$$

In particular ϕ_μ in the two gauges differ by $-\nabla_\mu(\beta/\beta_0)$ and since ϕ_μ is a part of Weyl geometry and the Dirac field represents the quantum mass one can conclude that the quantum effects are geometrized (on the other hand, Eq. (3.36) shows that ϕ_μ is not independent of β so the Weyl vector is determined by the quantum mass and thus the geometrical aspects of the manifold are related to quantum effects).

3.3 The Role of Quantum Potential in Cosmology

The quantum potential has the merit to provide a direct physical interpretation of quantum cosmology and thus a direct physical explanation of the behaviour of the universe as a whole. In particular, the Bohmian interpretation of Wheeler-de Witt equation gives the following relevant results that could be found in the literature:

- As Bohm's theory describes a single system, from in a Bohmian picture we have not the conceptual problem of the meaning of the universe's wave function in quantum cosmology.
- In Bohmian quantum cosmology the quantum force can remove the big bang singularity, because it can behave as a repulsive force [26, 27].
- The quantum force may be present in large scales because the quantum effects of quantum potential are independent of the scale [28].
- Bohmian quantum cosmology allows us to avoid any singularity—which characterizes instead super string cosmology—between inflation and the successive decelerating expansion: there is simply a smooth change from the two different stages [29].

- Real time tunneling can be occurred in the classically forbidden regions, through the quantum potential (for this effect in a closed de Sitter universe in $2+1$ dimensions see, for example, the Ref. [30]).

For the treatment of Wheeler-deWitt (WDW) equation in the context of a Bohmian framework, the discussion made here is based on the references [3, 12–14, 31] and takes into consideration also the papers [18, 19, 32–62].

The Lagrangian density for general relativity can be written in the form

$$L = \sqrt{-g}R = \sqrt{q}N\left({}^{(3)}R + Tr(K^2) \right) \tag{3.45}$$

where ${}^{(3)}R$ is the three-dimensional Ricci scalar, K_{ij} is the extrinsic curvature, N is the lapse function, q_{ij} is the induced spatial metric, $q = \det q_{ij}$. The canonical momentum of the 3-metric is given by

$$p^{ab} = \frac{\partial L}{\partial \dot{q}_{ab}} = \sqrt{q}\left(K^{ab} + q^{ab} Tr(K) \right). \tag{3.46}$$

The classical Hamiltonian is

$$H = \int d^3x \sqrt{q}(NC + N_i C^i) \tag{3.47}$$

where N_i is the shift function,

$$C = {}^{(3)}R + \frac{1}{q}\left(Tr(p^2) - \frac{1}{2}\left(Tr(p)^2 \right) \right) = -2G_{\mu\nu}n^\mu n^\nu \tag{3.48}$$

$$C^i = -2G_{\mu i}n^\mu, \tag{3.49}$$

n^μ being the normal vector to the spatial hypersurfaces given by $n^\mu = (1/N, -\vec{N}/N)$.

In a Bohmian approach to cosmology one has to add a quantum potential for the gravitational field (which emerges from WDW equation) to the classical Hamiltonian. The standard Wheeler-DeWitt (WDW) equation which characterizes the wave-functional Ψ of the universe can be written in the following regularized form (here we have made the position $\hbar = c = 1$):

$$\left[(8\pi G)G_{abcd}p^{ab}p^{cd} + \frac{1}{16\pi G}\sqrt{q}\left(2\Lambda - {}^{(3)}R \right) \right]\Psi = 0. \tag{3.50}$$

where $G_{abcd} = \frac{1}{2}\sqrt{q}(q_{ac}q_{bd} + q_{ad}q_{bc} - q_{ab}q_{cd})$ is the supermetric, p^{ab} are the momentum operators related to the 3-metric q_{ab}, Λ is the cosmological constant, G is the gravitational constant. The WDW equation has the following important features:

1. The time parameter which defines the foliation of the space-time, does not appear in it (we have thus the so-called time-problem in quantum gravity).

2. A different ordering of factors leads to a different result.
3. In practice, in order to solve the WDW equation, instead of using an infinite-dimensional superspace, we must limit ourselves to a mini-superspace in which some of the degrees of freedom are non-frozen.
4. It is necessary for the wave-function to be square-integrable, in order to have a probabilistic interpretation for it. But this is not possible for all cases, because a precise definition of the inner product is not known in quantum gravity.

Here our purpose is to analyse the Bohmian version of WDW equation in order to see what results it allows us to obtain about the geometry of space. In a Bohmian approach to WDW equation (3.), by decomposing the wave-functional Ψ in polar form $\Psi = R\,e^{iS/\hbar}$ one obtains a modified Hamilton–Jacobi equation

$$(8\pi G)G_{abcd}\frac{\delta S}{\delta q_{ab}}\frac{\delta S}{\delta q_{cd}} - \frac{1}{16\pi G}\sqrt{q}\left(2\Lambda - {}^{(3)}R\right) + Q_G = 0, \tag{3.51}$$

where

$$Q_G = \hbar^2 NqG_{abcd}\frac{1}{R}\frac{\delta^2 R}{\delta q_{ab}\delta q_{cd}}. \tag{3.52}$$

The term Q_G is the quantum potential for the gravitational field. Equation (3.51) indicates that the only difference between classical and quantum universes is the existence of the quantum potential in the second one. This means that, in order to obtain a quantum regime, one must modify the classical constraints via relation

$$C \to C + \frac{Q_G}{\sqrt{q}N}; C_i \to C_i. \tag{3.53}$$

As regards the constraint algebra one can use the integrated forms of the constraints defined as

$$C(N) = \int d^3x\sqrt{q}NC; \quad \tilde{C}(\vec{N}) = \int d^3x\sqrt{q}N^iC_i \tag{3.54}$$

which satisfy the following algebra

$$\{\tilde{C}(\vec{N}), \tilde{C}(\vec{N}')\} = \tilde{C}(\vec{N}\cdot\nabla\vec{N}' - \vec{N}'\cdot\nabla\vec{N}); \{\tilde{C}(\vec{N}), C(\vec{N})\}$$
$$= C(\vec{N}\cdot\nabla\vec{N}); \{C(N), C(N')\} \approx 0. \tag{3.55}$$

On the basis of Eq. (3.55), the first 3-diffeomorphism subalgebra and the second 3-diffeomorphism subalgebra do not change with respect to the classical situation; instead, in the third the quantum potential changes the Hamiltonian constraint algebra dramatically according to relation

$$\frac{1}{N}\frac{\delta}{\delta q_{ab}}\frac{Q}{\sqrt{q}} = \frac{3}{4\sqrt{q}}q_{cd}p^{ab}p^{cd}\delta(x-z) - \frac{\sqrt{q}}{2}q^{ab}\left({}^{(3)}R - 2\Lambda\right)\delta(x-z) - \sqrt{q}\frac{\delta^{(3)}R}{\delta q_{ab}} \tag{3.56}$$

giving a result weakly equal to zero for the Poisson bracket (namely zero when the equations of motion are satisfied).

The next step is to develop the quantum Einstein equations inside the geometry of space represented by the algebra here mentioned. In this regard, we follow the references [3, 4, 12, 63, 64] of F. Shojai and A. Shojai. Let us consider the quantum Einstein equations in absence of source of matter-energy. For the dynamical parts, by considering the Hamilton equations, one arrives to equation

$$G^{ab} = -\frac{1}{N}\frac{\delta \int d^3 x Q_G}{\delta q_{ab}} \tag{3.57}$$

which means that the quantum force modifies the dynamical parts of Einstein's equations (in Eq. (3.57), $G^{\mu\nu} = R^{\mu\nu} - \frac{1}{2}q^{\mu\nu}R$ is, of course, Einstein's tensor). For the non-dynamical parts, by using the constraint relations (3.48) and (3.50), one obtains

$$G^{00} = \frac{Q_G}{2N^3\sqrt{q}}; G^{0i} = -\frac{Q_G}{2N^3\sqrt{q}}N^i. \tag{3.58}$$

Equation (3.58) can also be written as

$$G^{0\mu} = \frac{Q_G}{2\sqrt{-g}}g^{0\mu} \tag{3.59}$$

which shows that the non-dynamical parts are modified by the quantum potential. It is interesting to observe that such modified Einstein's equations are covariant under spatial and temporal diffeomorphisms.

By inserting the matter quantum potential and by introducing the energy–momentum tensor in these equations, one obtains:

$$G^{ab} = -kT^{ab} - \frac{1}{N}\frac{\delta \int d^3 x (Q_G + Q_m)}{\delta q_{ab}}; \quad G^{0\mu} = -kT^{0\mu} - \frac{1}{N}\frac{(Q_G + Q_m)}{2\sqrt{-q}}q^{0\mu} \tag{3.60}$$

where

$$Q_m = \hbar^2 \frac{N\sqrt{H}}{2}\frac{\delta^2 R}{\delta\varphi^2} \tag{3.61}$$

is the matter quantum potential (φ being the matter field), and, of course,

$$Q_G = \hbar^2 N q G_{abcd}\frac{1}{R}\frac{\delta^2 R}{\delta q_{ab}\delta q_{cd}} \tag{3.52}$$

is the usual quantum potential for the gravitational field.

Therefore, the Bohmian approach to WDW equation based on Eq. (3.60) leads to general Bohm-Einstein equations of the form (3.51) which are in fact the quantum version of Einstein's equations. Here, since regularization only influences the quantum potential (cfr. also [4, 5, 12, 63, 64]), for any regularization the Bohm-Einstein equations (3.60) are the same, and are invariant under temporal $\otimes$ spatial diffeomorphisms; so, they can be written also in the equivalent compact form

$$G^{\mu\nu} = -kT^{\mu\nu} + S^{\mu\nu} \qquad (3.62)$$

where

$$S^{0\mu} = \frac{Q_G + Q_m}{2\sqrt{-g}} g^{0\mu}; S^{ab} = -\frac{1}{N} \frac{\delta \int d^3x (Q_G + Q_m)}{\delta g_{ab}}. \qquad (3.63)$$

$S^{\mu\nu}$ is the quantum corrector tensor, under the temporal $\otimes$ spatial diffeomorphisms subgroup of the general coordinate transformations. One can say that Eq. (3.60) (and the equivalent Eq. (3.62)) describe the geometry of physical space which is derived from WDW equation and that the quantum corrector tensor $S^{\mu\nu}$ (defined by Eq. (3.63) and determined by the matter quantum potential and by the quantum potential for the gravitational field) indicates the change, the deformation of the geometry of the physical space produced by matter and gravity in WDW equation's regime.

In synthesis, according to the model developed by F. Shoiai and A. Shojai, in the Bohmian approach to WDW equation the complete set of equations to be solved in order to describe and obtain the geometry of physical space in quantum cosmology is constituted by: the Eq. (3.62), the WDW equation and the appropriate equation of matter field given by matter Lagrangian. It must be noted here that solving the above mentioned equations is mathematically equivalent to solving the WDW equation and then using its decomposition to Hamilton–Jacobi equation and continuity equation, and extracting the Bohmian trajectories (cfr. also [16–19], Ref. [8] in Chap. 1 for further details in this regard). It is also interesting to remark that the quantum Einstein's equations have also been derived for the Robertson–Walker metric in the papers [65, 66], but there neither any try is done to write the equations for a general metric, nor the symmetries are investigated. Finally, in this Bohmian model for WDW equation suggested by F. Shojai and A. Shojai, by getting the divergence of Eq. (3.62), one obtains:

$$\nabla_\mu T^{\mu\nu} = \frac{1}{k} \nabla_\mu S^{\mu\nu} \qquad (3.69)$$

which can be interpreted as an energy conservation law for WDW equation.

In the Bohmian approach to quantum cosmology, the fundamental object of quantum gravity, the geometry of 3-dimensional spacelike hypersurfaces, is supposed to exist independently on any observation or measurement, as well as its canonical momentum, the extrinsic curvature of the spacelike hypersurfaces. Its evolution, labeled by some time parameter, is dictated by a quantum evolution that is different from the classical one due to the presence of a quantum potential which appears naturally from the WDW equation. The quantum potential, by introducing appropriate corrector terms in Einstein's equations, determines a deformation of the geometry of the physical space with respect to the classical situation.

Bohm's approach to WDW equation has been indeed applied to many mini-superspace models obtained by the imposition of homogeneity of the spacelike hypersurfaces. In this regard, the reader can find details, for example, in the references [21, 23, 66–69]. Here, the classical limit, the singularity problem, the cosmological constant problem and the time issue have been discussed. For instance, in some of these papers it was shown that in models involving scalar fields or radiation, which are good representatives of the matter content of the early universe, the singularity can be clearly avoided by quantum effects. In the de Broglie-Bohm interpretation of quantum cosmology, the quantum potential becomes important near the singularity, yielding a repulsive quantum force counteracting the gravitational field, avoiding the singularity and yielding inflation.

Other interesting results about the geometry of space determined by the quantum potential emerging from WDW equation have been obtained by Pinto-Neto and are summarized in his papers [33, 70–75]. As regards the Bohmian approach to quantum cosmology in the case of homogeneous minisuperspace models, Pinto Neto showed that there is no problem of time and that quantum effects can avoid the initial singularity, create inflation, and isotropize the Universe. As regards the general case of superspace canonical quantum cosmology, it has been demonstrated that the Bohmian evolution of the 3-geometries, irrespective of any regularization and factor ordering of the WDW equation, can be obtained from a specific hamiltonian, which is of course different from the classical one.

By following Pinto-Neto's treatment, the Bohm-de Broglie interpretation of canonical quantum cosmology leads to a quantum geometrodynamical picture where the Bohmian quantum evolution of three geometries may form, depending on the wave functional, a consistent non degenerate four-geometry (which can be euclidean—for a very special local form of the quantum potential—or hyperbolic), and a consistent but degenerate four-geometry indicating the presence of special vector fields and the breaking of the space-time structure as a single entity (in a wider class of possibilities).

Pinto-Neto starts by writing the WDW equation in the following unregulated form in the coordinate representation

$$\left[-\hbar^2\left(kG_{abcd}\frac{\delta}{\delta q_{ab}}\frac{\delta}{\delta q_{cd}} + \frac{1}{2\sqrt{q}}\frac{\delta^2}{\delta\phi^2}\right) + V\right]\Psi = 0 \qquad (3.70)$$

where V is the classical potential given by relation

$$V = \sqrt{q}\left[-\frac{1}{k}\left(^{(3)}R - 2\Lambda\right) + \frac{1}{2}q^{ab}\partial_a\phi\partial_b\phi + U(\phi)\right] \qquad (3.71)$$

and there is the constraint $-2q_{ab}\nabla_b\left(\frac{\delta\Psi}{\delta q_{ab}}\right) + \left(\frac{\delta\Psi}{\delta\phi}\right)\partial_a\phi = 0$. Writing the wave functional in polar form, Eq. (3.70) yields

$$kG_{abcd}\frac{\delta S}{\delta q_{ab}}\frac{\delta S}{\delta q_{cd}}+\frac{1}{2\sqrt{q}}\left(\frac{\delta S}{\delta\phi}\right)^2+V+Q=0 \qquad (3.72)$$

where the quantum potential is given by

$$Q=-\frac{\hbar^2}{R}\left(kG_{abcd}\frac{\delta^2 R}{\delta q_{ab}\delta q_{cd}}+\frac{1}{2\sqrt{q}}\frac{\delta^2 R}{\delta\phi^2}\right) \qquad (3.73)$$

In Pinto-Neto's model, a non degenerate four-geometry can be attained if the quantum potential (3.73) assumes the specific form

$$Q=-\sqrt{q}\left[(\varepsilon+1)\left(-\frac{1}{k}{}^{(3)}R+\frac{1}{2}q^{ab}\partial_a\phi\partial_b\phi\right)+\frac{2}{k}\left(\varepsilon\overline{\Lambda}+\Lambda\right)+\varepsilon\overline{U}(\phi)+U(\phi)\right]$$

$$(3.74)$$

(where ε is a constant which can be ± 1 depending if the four-geometry in which the 3-geometries are embedded is Euclidean or hyperbolic, providing thus the conditions for the existence of spacetime). If $\varepsilon=-1$, condition in which the spacetime is hyperbolic, the quantum potential (3.74) becomes

$$Q=-\sqrt{q}\left[\frac{2}{k}\left(\varepsilon\overline{\Lambda}+\Lambda\right)-\varepsilon\overline{U}(\phi)+U(\phi)\right] \qquad (3.75)$$

and thus is like a classical potential. The effect of the quantum potential (3.75) is to renormalize the cosmological constant and the classical scalar field potential, nothing more. In this domain, the quantum geometrodynamics turns out to be indistinguishable from the classical one. If $\varepsilon=+1$, condition in which the spacetime is Euclidean, the quantum potential (3.74) becomes

$$Q=-\sqrt{q}\left[2\left(-\frac{1}{k}{}^{(3)}R+\frac{1}{2}q^{ab}\partial_a\phi\partial_b\phi\right)+\frac{2}{k}\left(\varepsilon\overline{\Lambda}+\Lambda\right)+\varepsilon\overline{U}(\phi)+U(\phi)\right] \quad (3.76)$$

In this case, the quantum potential not only renormalizes the cosmological constant and the classical scalar field potential but also changes the signature of spacetime. The total potential $V+Q$ corresponding to $\varepsilon=+1$ may describe some era of the early universe when it had Euclidean signature, but not the present era, when it is hyperbolic. The transition between these two phases occurs in a hypersurface where $Q=0$, which is the classical limit. On the basis of Pinto-Neto's treatment, one can conclude that if a quantum space-time exists with different features with respect to the classical observed one, then it must be euclidean. In other words, the only relevant quantum effect which maintains the non-degenerate nature of the four-geometry of space-time is its change of signature to a euclidean one. The other quantum effects are either irrelevant or break completely the spacetime structure.

However, as shown by Pinto-Neto, the three geometries evolving under the influence of a quantum potential do not in general stick together to form a non degenerate four-geometry, a single space-time with the causal structure of

relativity. Among the consistent Bohmian evolutions, the most general structures
that are formed are degenerate four-geometries with alternative causal structures.
In the case of consistent quantum evolution with degenerate four-geometry,
Pinto-Neto showed that any real solution of the WDW equation leads to a
structure which is the idealization of the strong gravity limit of general relativity.
An example of this situation is obtained, for real solutions of WDW equation, if
the phase S is zero and thus $Q = -V$ which implies that the quantum super
Hamiltonian contains only the kinetic term. Another example of this situation
occurs for $Q = \gamma V$ where γ is a (non-local) function of the functional S. In this
peculiar case, one finds that a breaking of space-time emerges from the non-local
character of the quantum potential. This type of geometry, the so-called degenerate four-geometry has already been studied also in the Ref. [76] and might be
considered as the correct quantum geometrodynamical description of the young
universe. It would be also interesting to investigate if these structures have a
classical limit leading to the usual four-geometry of classical cosmology.

Moreover, for non-local quantum potentials, Pinto-Neto showed that apparently
inconsistent quantum evolutions are in fact consistent if restricted to the Bohmian
trajectories satisfying the guidance relations corresponding to the Bohm-Einstein
Eq. (3.62). If one wants to be strict and impose that quantum geometrodynamics
does not break spacetime by determining a non degenerate four-geometry, then
one will have stringent boundary conditions, like the form (3.74) for the quantum
potential (which implies a severe restriction on the solutions of the WDW
equation).

It is also important to underline that Pinto-Neto's results, although were
obtained without assuming any particular factor ordering and regularization of the
WDW equation and can draw a lot of information using the quantum potential
approach (without appealing to any probabilistic notion), nonetheless are limited
by many strong assumptions tacitly made, as supposing that a continuous three-
geometry exists at the quantum level (quantum effects could also destroy it), or the
validity of quantization of standard general relativity, forgetting other developments like, for example, string theory.

To conclude this chapter, we focus our attention on the papers of G. D. Barbosa
and N. Pinto-Neto [77–80] where relevant results are obtained as regards non-
commutative geometry and cosmology in connection with the quantum potential in
the picture of a Kantowski-Sachs universe. In these papers Barbosa and Pinto-Neto
take under consideration a WDW equation of the form

$$\left(G^{ijkl} \frac{\delta}{\delta q^{ij}} \frac{\delta}{\delta q^{kl}} + q^{1/2} R^{(3)} \right) \Psi \left[q^{ij} \right] = 0 \tag{3.77}$$

If one puts $\Psi = R\, e^{iS}$ in (3.77) one obtains

$$G^{ijkl} \frac{\delta S}{\delta q^{ij}} \frac{\delta S}{\delta q^{kl}} - q^{1/2} R^{(3)} + Q = 0 \tag{3.78}$$

and

$$G^{ijkl}\frac{\delta S}{\delta q^{ij}}\left(A^2\frac{\delta S}{\delta q^{kl}}\right)=0 \qquad (3.79)$$

where

$$Q=-\frac{1}{A}G^{ijkl}\frac{\delta^2 A}{\delta q^{ij}\delta q^{kl}} \qquad (3.80)$$

is the quantum potential.

In the model of a Kantowski-Sachs universe the line element in the Misner parametrization is

$$ds^2=-N^2dt^2+e^{2\sqrt{3}\beta}dr^2+e^{-2\sqrt{3}\beta}e^{2\sqrt{3}\Omega}\left(d\theta^2+\sin^2(\theta)d\varphi^2\right). \qquad (3.81)$$

In order to distinguish the role of the quantum and noncommutative effects in a non-commutative quantum universe one starts now with a Kantowski-Sachs geometry in the commutative classical version, then a commutative quantum model is introduced in a minisuperspace approach and a non-commutative quantum picture is considered by taking the canonical commutation relation

$$\left[\hat{\Omega},\hat{\beta}\right]=i\theta. \qquad (3.82)$$

According to the Weyl quantization procedure the realization of (3.82) in terms of commutative functions is made by the Moyal star product defined via equation

$$f(\Omega_c,\beta_c)*g(\Omega_c,\beta_c)=f(\Omega_c,\beta_c)e^{i(\theta/2)\left(\overleftarrow{\partial}_{\Omega_c}\overrightarrow{\partial}_{\beta_c}-\overleftarrow{\partial}_{\beta_c}\overrightarrow{\partial}_{\Omega}\right)}g(\Omega_c,\beta_c) \qquad (3.83)$$

where Ω_c and β_c are commutative coordinates which are called Weyl symbols of the operators $\hat{\Omega}$ and $\hat{\beta}$. In order to compare evolutions with the same time parameter as above one here fixes the gauge $N=24\exp(-\sqrt{3}\beta-2\sqrt{3}\Omega)$ and then the WDW Eq. (3.77) for the non-commutative Kantowski-Sachs model becomes of the following Klein-Gordon form

$$\left[-P^2_{\Omega_c}+P^2_{\beta_c}-48e^{-2\sqrt{3}\Omega_c}\right]*\Psi(\Omega_c,\beta_c)=0. \qquad (3.84)$$

By using properties of the Moyal bracket one can write the potential term as

$$V(\Omega_c,\beta_c)*\Psi(\Omega_c,\beta_c)=V\left(\Omega_c+i\frac{\theta}{2}\partial_{\beta_c},\beta_c-i\frac{\theta}{2}\partial_{\Omega_c}\right)\Psi(\Omega_c,\beta_c)$$
$$=V\left(\hat{\Omega},\hat{\beta}\right)\Psi(\Omega_c,\beta_c) \qquad (3.85)$$

where

$$\hat{\Omega} = \hat{\Omega}_c - \frac{\theta}{2}\hat{P}_{\beta_c}; \; \hat{\beta} = \hat{\beta}_c + \frac{\theta}{2}\hat{P}_{\Omega_c}. \tag{3.86}$$

The WDW equation then reads as

$$\left[-\hat{P}_{\Omega_c}^2 + \hat{P}_{\beta_c}^2 - 48 e^{-2\sqrt{3}\hat{\Omega}_c + \sqrt{3}\theta\hat{P}_{\beta_c}} \right] \Psi\left(\hat{\Omega}_c, \hat{\beta}_c\right) = 0. \tag{3.87}$$

From these equations, two consistent interpretations for the cosmology are possible. The first lies in considering the Weyl symbols Ω_c and β_c as the constituents of the physical metric: this interpretation implies that things can be made essentially commutative with a modified interaction. The second, as adopted in the papers [77, 79, 80], is based on the idea that the Weyl symbols represent auxiliary coordinates, and thereby implies the perspective to study the evolution of a non-commutative quantum universe.

In Bohmian non-commutative quantum cosmology developed by Pinto-Neto and Barbosa in the papers [77–80], the metric variables can be described as well defined entities, although the operators $\hat{\Omega}$ and $\hat{\beta}$ satisfy Eq. (3.82). Here, one can give an objective meaning to the wavefunction and the metric variables Ω and β and the wave function is obtained by solving Eq. (3.84). The evolution law for Ω and β is obtained by associating a function $U(\Omega_c, \beta_c)$ to $\hat{A}\left(\hat{\Omega}_c, \hat{\beta}_c, \hat{P}_{\Omega_c}, \hat{P}_{\beta_c}\right)$ according to the rule

$$B[\hat{A}] = \frac{\text{Re}\left[\Psi^*(\Omega_c, \beta_c)\hat{A}\left(\Omega_c, \beta_c, -i\hbar\partial_{\Omega_c}, -i\hbar\partial_{\beta_c}\right)\Psi(\Omega_c, \beta_c)\right]}{\Psi^*(\Omega_c, \beta_c)\Psi(\Omega_c, \beta_c)} = U(\Omega_c, \beta_c) \tag{3.88}$$

where the real part takes into account the hermiticity of $\hat{A}$ (the B here refers to the idea of "beable"). Applying this to the operators $\hat{\Omega}$ and $\hat{\beta}$ one obtains the following relations

$$\Omega(\Omega_c, \beta_c) = B[\hat{\Omega}] = \frac{\text{Re}\left[\Psi^*(\Omega_c, \beta_c)\hat{\Omega}(\Omega_c, -i\hbar\partial_{\beta_c})\Psi(\Omega_c, \beta_c)\right]}{\Psi^*(\Omega_c, \beta_c)\Psi(\Omega_c, \beta_c)} = \Omega_c - \frac{\theta}{2}\partial_{\beta_c}S \tag{3.89}$$

$$\beta(\Omega_c, \beta_c) = B[\hat{\beta}] = \frac{\text{Re}\left[\Psi^*(\Omega_c, \beta_c)\hat{\beta}(\beta_c, -i\hbar\partial_{\Omega_c})\Psi(\Omega_c, \beta_c)\right]}{\Psi^*(\Omega_c, \beta_c)\Psi(\Omega_c, \beta_c)} = \beta_c + \frac{\theta}{2}\partial_{\Omega_c}S. \tag{3.90}$$

Thus the relevant information for universe evolution can be extracted from the guiding wave $\Psi(\Omega_c, \beta_c)$ by first computing the associated canonical position tracks $\Omega_c(t)$ and $\beta_c(t)$. Then one obtains $\Omega(t)$ and $\beta(t)$ by evaluating Eqs. (3.89) and (3.90) at $\Omega_c = \Omega_c(t)$ and $\beta_c = \beta_c(t)$. Differential equations for the canonical positions $\Omega_c(t)$ and $\beta_c(t)$ can be found by identifying $\dot{\Omega}_c(t)$ and $\dot{\beta}_c(t)$ with the beables associated with their time evolution and formulas are worked out in [62–64].

The combination of non-commutative geometry and Bohmian quantum cosmology in the Kantowski-Sachs universe, which can be derived from the works [78–80] by Pinto-Neto and Barbosa can be considered as a fusion of two apparently opposite ways of thinking, one fuzzy and the other referring ontologically to point particles. Every Hermitian operator can be associated with an ontological element and by averaging the beable $B[\hat{A}]$ over an ensemble of particles with probability density $\rho = |\Psi|^2$ one gets the same result as computing the expectation value of the observable $\hat{A}$ in standard operational formalism. In the Kantowski-Sachs universe, where WDW is of Klein-Gordon type (with no notion of probability) the beable mapping is well defined, even in the non-commutative case. In the commutative context this formulation leads to the Bohmian quantum gravity proposed by Holland in Ref. [8] in Chap. 1 in the minisuperspace approximation. The work by Pinto-Neto and Barbosa here mentioned shows that noncommutativity can modify appreciably the universe evolution in the quantum context (qualitatively as well as quantivately).

The non-commutative geometry of quantum cosmology in a Kantowski-Sachs universe described by Eqs. (3.81)–(3.90) can be considered as an attempt to develop a fundamental level of reality, namely the implicate order, which determines the action of the quantum potential for WDW Eq. (3.80). Equations (3.81–3.90) can be considered, in the context of quantum cosmology—in the picture of a Kantowski-Sachs universe—as a sort of counterparts of Eqs. (1.149) and (1.153) or (1.171) and (1.173) which describe the non-commutative underlying process of non-relativistic quantum phenomena. According to Hiley's view, in Pinto-Neto's and Barbosa's quantum cosmology the non-commutative algebra in the context of the Kantowski-Sachs universe can be considered as the fundamental structure and the quantum potential for WDW equation emerges directly from it. The quantum potential applied to Cosmology seems to suggest a "Big Bang" interpretation as a sort of nucleation by vacuum, idea which has been developed by G. Chapline [81] and later by I. Licata and L.Chiatti [82, 83], using the Bohm's concept of holomovement.

References

1. Carroll, R.W.: Fluctuations, Information, Gravity and the Quantum Potential. Springer, Dordrecht (2006)
2. Shojai, A.: Quantum, gravity and geometry. Int. J. Mod. Phys. A **15**, 1757 (2000)
3. Shojai, F., Shojai, A.: Pure quantum solutions of Bohmian quantum gravity. Jour. High Energy Physics **05**, 037 (2001)
4. Shojai, F., Shojai, A.: Quantum Einstein's equations and constraints algebra. Pramana **58**(1), 13–19 (2002)
5. Shojai, F., Shojai, A.: On the relation of Weyl geometry and and Bohmian quantum gravity. Gravitation Cosmol. **9**(3), 163–168 (2003)
6. Shojai, F., Shojai, A.: Constraints Algebra and equations of motion in Bohmian interpretation of quantum gravity. Classical Quant. Grav. **21**, 1–9 (2004)

7. Shojai, F., Shojai, A.: Non-minimal scalar-tensor theories and quantum gravity. Inter. Jour. Mod. Phys. **A15**(13), 1859–1868 (2000)
8. Shojai, A., Golshani, M.: Direct particle interaction as the origin of the quantal behaviours. arXiv:quant-ph/9812019 (1996)
9. Shojai, A., Golshani, M.: On the relativistic quantum force. arXiv:quant-ph/9612023 (1996)
10. Shojai, A., Golshani, M.: Some observable results of the retarded's Bohm's theory. arXiv:quant-ph/9612020 (1996)
11. Shojai, A., Golshani, M.: Is superluminal motion in relativistic Bohm's theory observable? arXiv:quant-ph/9612021 (1996)
12. Shojai, F., Shojai, A.: Constraint algebra and the equations of motion in the Bohmian interpretation of quantum gravity. Class. Quant. Grav. **21**, 1–9 (2004)
13. Shojai, F., Shojai, A.: Causal loop quantum gravity and cosmological solutions. Europhys. Lett. **71**(6), 886–892 (2005)
14. Shojai. F., Shojai, A. (2005). Constraint algebra in causal loop quantum gravity. In : Rahvar, S., Sadooghi, N., Shojai F (ed.) Proceedings of the XI Regional Conference, pp. 114–116. Tehran, Iran. World Scientific, 3–6 May 2004
15. de-Broglie, L.: Non-linear Wave Mechanics (translated by A.J. Knodel). Elsevier Publishing Company, Amsterdam (1960)
16. Horiguchi, T.: Quantum potential interpretation of the Wheeler-deWitt equation. Mod. Phys. Lett. A **9**(16), 1429–1443 (1994)
17. Blaut, A., Glikman, J.K.: Quantum potential approach to class of quantum cosmological models. Class. Quant. Grav. **13**, 39–50 (1996)
18. Kim, S.P.: Quantum potential and cosmological singularities. Phys. Lett. A **236**(1/2), 11–15 (1997)
19. Kim, S.P.: Problem of unitarity and quantum corrections in semi-classical quantum gravity. Phys. Rev. D **55**(12), 7511–7517 (1997)
20. Cushing, J.T.: The causal quantum theory program. In: Cushing, J.T., Fine, A., Goldstein, S. (eds.) Bohmiam Mechanics and Quantum Theory: An Appraisal. Kluwer Academic Publishers, Boston (1996)
21. Shojai, F., Shojai, A., Golshani, M.: Conformal transformations and quantum gravity. Mod. Phys. Lett. A **13**(34), 2725–2729 (1998)
22. Shojai, F., Shojai, A., Golshani, M.: Scalar-tensor theories and quantum gravity. Mod. Phys. Lett. A **13**(36), 2915–2922 (1998)
23. Shojai, A., Shojai, F., Golshani, M.: Nonlocal effects in quantum gravity. Mod. Phys. Lett. A **13**(37), 2965–2969 (1998)
24. Shojai, F., Shojai, A.: On the relation of Weyl geometry and Bohmian quantum mechanics. Gravitation Cosmol. **9**, 163–168 (2003)
25. Cho, Y.M., Park, D.H.: Higher-dimensional unification and fifth force. Il Nuovo Cimento B **105**(8/9), 817–829 (1990)
26. de Barros, J.A., Pinto-Neto, N., Sagioro-Leal, M.A.: The causal interpretation of dust and radiation fluid non-singular quantum cosmologies. Phys. Lett. A **241**(4), 229–239 (1998)
27. Shojai, F., Golshani, M.: On the geometrization of Bohmian mechanics: a new suggested approach to quantum gravity. Int. J. Mod. Phys. A **13**(4), 677–693 (1998)
28. Colistete, J.R., Fabris, J.C., Pinto–Neto, N.: Singularities and classical limit in quantum cosmology with scalar fields. Phys. Rev. D **57**(8), 4707–4717 (1998)
29. Pinto-Neto, N., Colistete, R.: Graceful exit from inflation using quantum cosmology. Phys. Lett. A **290**(5/6), 219–226 (2001)
30. Kenmoku, M., Sato, R., Uchida, S.: Classical and quantum solutions of (2 + 1)-dimensional gravity under the de Broglie-Bohm interpretation. Class. Quantum Grav. **19**(4), 799 (2002)
31. Pinto-Neto, N.: Quantum cosmology: how to interpret and obtain results. Braz. J. Phys. **30**(2), 330–345 (2000)
32. Ashtekar, A., Lewandowski, J.: Quantum field theory of geometry. arXiv:hep-th/9603083 (1996)

33. Ashtekar, A., Corichi, A., Zapata, J.: Quantum theory of geometry III: non-commutativity of Riemannian structures. Class. Quant. Grav. **15**, 2955–2972 (1998)
34. Ashtekar, A., Lewandowski, J.: Background independent quantum gravity: a status report. Class. Quant. Grav. **21**, 53–152 (2004)
35. Ashtekar, A., Isham, C.: Representations of holonomy algebras of gravity and non-abelian gauge theories. Class. Quantum Grav. **9**, 1433–1467 (1992)
36. Ashtekar, A.: Gravity and the quantum. New J. Phys. **7**, 198 (2005)
37. Baez, J. (ed.): Knots and Quantum Gravity. Oxford University Press, Oxford (1994)
38. Biswas, S., Shaw, A., Biswas, D.: Schrödinger Wheeler-DeWitt equation in multidimensional cosmology. Int. J. Mod. Phys. D **10**, 585–594 (2001)
39. Dittrich, B., Thiemann, T.: Testing the master constraint programme for loop quantum gravity I. General framework. Class. Quant. Grav. **23**, 1025–1046 (2006)
40. Dittrich, B., Thiemann, T.: Testing the master constraint programme for loop quantum gravity II. Finite dimensional systems. Class. Quant. Grav. **23**, 1067–1088 (2006)
41. Dittrich, B., Thiemann, T.: Testing the master constraint programme for loop quantum gravity III. SL(2R) models. Class. Quant. Grav. **23**, 1089–1120 (2006)
42. Dittrich, B., Thiemann, T.: Testing the master constraint programme for loop quantum gravity IV. Free field theories. Class. Quant. Grav. **23**, 1121–1142 (2006)
43. Dittrich, B., Thiemann, T.: Testing the master constraint programme for loop quantum gravity V. Interaction field theories. Class. Quant. Grav. **23**, 1143–1162 (2006)
44. Gambini, R., Pullin, J.: Loops, Knots, Gauge Theories and Quantum Gravity. Cambridge University Press, Cambridge (2000)
45. Horiguchi, T., Maeda, K., Sakamoto, M.: Analysis of the Wheeler-DeWitt equation beyond Planck scale and dimensional reduction. Phys. Lett. **B344**, 105–109 (1995)
46. Kenmoku, M., Kubortani, H., Takasugi, E., Yamazaki, Y.: De Broglie-Bohm interpretation for analytic solutions of the Wheeler-DeWitt equation in spherically symmetric space-time. Progress Theoret. Phys. **105**(6), 897–914 (2001)
47. Kiefer, C.: Quantum Gravity. Oxford University Press, Oxford (2004)
48. Gambini, R., Pullin, J.: A First Course in Loop Quantum Gravity. Oxford University Press, Oxford (2011)
49. Markopoulou, F., Smolin, L.: Quantum theory from quantum gravity. Phys. Rev. **D70**, 124029 (2004)
50. Moss, I.: Quantum Theory, Black Holes, and Inflation. Wiley, New York (1996)
51. Rovelli, C.: Quantum Gravity. Cambridge University Press, Cambridge (2004)
52. Rovelli, C.: Covariant hamiltonian formalism for field theory: Hamilton-Jacobi equation on the space G. arXiv:gr-qc/0207043 (2002)
53. Smolin, L.: Three Roads to Quantum Gravity. Oxford University Press, Oxford (2002)
54. Smolin, L.: An invitation to loop quantum gravity. arXiv:hep-th/0408048 (2004)
55. Smolin, L.: Matrix models as non-local hidden variables theories. arXiv:hep-th/0201031 (2002)
56. Thiemann, T.: Modern Canonical Quantum General Relativity. Cambridge Monographs on Mathematical Physics, Cambridge University Press, Cambridge (2008)
57. Thiemann, T.: Lectures on loop quantum gravity. Lect. Notes Phys. **631**, 41–135 (2003)
58. Thiemann, T.: Anomaly-free formulation of non-perturbative, four-dimensional. Lorentzian quantum gravity. Phys. Lett. B **380**, 257–264 (1996)
59. Thiemann, T.: Quantum spin dynamics. Class. Quant. Grav. **15**, 839–873 (1998)
60. Thiemann, T.: Quantum spin dynamics (qsd). 2. Class. Quant. Grav. **15**, 875–905 (1998)
61. Thiemann, T.: QSD IV: 2 + 1 Euclidean quantum gravity as a model to test. 3 + 1 Lorentzian quantum gravity. Class. Quant. Grav. **15**, 1249–1280 (1998)
62. Thiemann, T.: QSD 5: Quantum gravity as the natural regulator of matter quantum field teories. Class. Quant. Grav. **15**, 1281–1314 (1998)
63. Shojai, A., Shojai, F.: About some problems raised by the relativistic form of de Broglie-Bohm theory of pilot wave. Phys. Scr. **64**, 413 (2001)

64. Shojai, A., Shojai, F., Naresh, D.: Static Einstein's universe as a quantum solution of causal quantum gravity. Int. Jour. of Mod. Phys. A **20**(13), 2773–2780 (2005)
65. Vink, J.: Quantum mechanics in terms of discrete beables. Phys. Rev. A **48**(3), 1808–1818 (1993)
66. Vink, J.: Quantum potential interpretation of the wave function of the universe. Nucl. Phys. B **369**(3), 707–728 (1992)
67. de Barros, J.A., Pinto-Neto, N.: The causal interpretation of quantum mechanics and the singularity problem and time issue in quantum cosmology. Int. J. of Mod. Phys. D **7**, 201 (1998)
68. Kowalski-Glikman, J., Vink, J.C.: Gravity-matter mini-superspace: quantum regime, classical regime and in between. Class. Quantum Grav. **7**, 901–918 (1990)
69. Squires, E.J.: A quantum solution to a cosmological Mystery. Phys. Lett. A **162**, 35 (1992)
70. Pinto-Neto, N.: The Bohm interpretation of quantum cosmology. Found. Phys. **35**(4), 577–603 (2005)
71. Pinto-Neto, N.: Evolution of perturbations in bouncing cosmological models. AIP Conf. Proc. **782**, 277–288 (2004)
72. Pinto-Neto, N., Santini, S.E.: The consistency of causal quantum geometrodynamics and quantum field theory. Gen. Rel. Grav. **34**(4), 505–532 (2002)
73. Pinto-Neto, N., Santini, S.E.: The accelerated expansion of the universe as a quantum cosmological effect. Phys. Lett. A **315**(1/2), 36–50 (2003)
74. Pinto-Neto, N., Fabris, J.C.: Quantum cosmology from the de Broglie-Bohm perspective. arXiv:1306.0820v1 [gr-qc] (2013)
75. Pinto-Neto, N., Santini, S.E.: Must quantum spacetimes be Euclidean? Phys. Rev. D **59**(12), 123517 (1999)
76. Dautcourt, G.: On the ultrarelativistic limit of general relativity. Acta Phys. Pol., B **29**(4), 1047–1055 (1998)
77. Barbosa, G., Pinto-Neto, N.: Noncommutative quantum mechanics and Bohm's ontological interpretation. Physical Review D **69**(6), 065014 (2004)
78. Barbosa, G., Pinto-Neto, N.: Noncommutative geometry and cosmology. Phys. Rev. D **70**, 103512 (2004)
79. Barbosa, G.: Noncommutative conformally coupled scalar field cosmology and its commutative counterpart. Phys. Rev. D **71**(6), 063511 (2005)
80. Barbosa, G.: On the meaning of the string-inspired noncommutativity and its implications. Jour. High Energy Phys. **5**, 24 (2003)
81. Chapline, G.: Geometrization of gauge fields. Quantum Theory and Gravitation, pp. 177–186. Academic Press, A.R. Marlow (1980)
82. Licata, I., Chiatti, L.: The archaic universe: big-bang, cosmological term and the quantum origin of time in projective relativity. Int. Jour. Theor. Phys. **48**, 1003–1018 (2009)
83. Licata, I., Chiatti, L.: Archaic universe and cosmological model: "big-bang" as nucleation by vacuum. Int. J. Theor. Phys. **49**, 2379–2402 (2010)

Chapter 4
Entropy, Information, Chaos and the Quantum Potential

4.1 A Short Survey About the Approaches to Quantum Entropy and Quantum Information

According to the modern research, several important results have been obtained as regards quantum entropy and quantum information. In this chapter we analyse the several approaches to quantum entropy (von Neumann, Kak, Sbitnev) and finally a geometric approach suggested recently by the authors of this book.

4.1.1 Von Neumann Entropy and Shannon Information

As is known, if one considers a quantum system characterized by the density operator ρ, the average information the experimenter obtains in repeated observations of many copies of an identically prepared mixed state is given by the von Neumann entropy

$$S_n(\rho) = \sum_{x=1}^{N} \lambda_x \ln \lambda_x, \qquad (4.1)$$

where λ_x are the eigenvalues of the density matrix associated with the system. From the von Neumann entropy (4.1) one can define the Shannon information

$$I = -S_n(\rho) \qquad (4.2)$$

(which takes the values $0 \leq I \leq \ln N$) which measures the average uncertainty of the events A_x corresponding with the N possible outcomes of an experiment and at the same time quantifies the accessible information. Thus, if one identifies the A_x labels for discrete states of a system, Eq. (4.2) can be interpreted as a measure of uncertainty of the state before this one is chosen and the Shannon entropy can be seen as a measure of the degree of ignorance concerning which possibility

I. Licata and D. Fiscaletti, *Quantum potential: Physics, Geometry and Algebra*, SpringerBriefs in Physics, DOI: 10.1007/978-3-319-00333-7_4, © The Author(s) 2014

(event A_x) may hold true in the set of all A_x with a given a priori probability distribution (Ref. [1] in Chap. 3).

In a similar way, for a random variable X with probability density $p(x)$, the information of X is defined as

$$I(X) = -\int dx p(x) \ln(p(x)). \qquad (4.3)$$

In the case of a phase space distribution of two variables P and Q having the form $\mu(p, q) = \langle z|\rho|z\rangle$ where $|z\rangle$ denotes coherent states, the Shannon information (4.3) may be generalized as

$$I(P, Q) = -\int \frac{dp dq}{2\pi\hbar} \mu(p, q) \ln(\mu(p, q)). \qquad (4.4)$$

As regards the information (4.4), in the paper [1] it has been shown that a lower bound is achieved in the form $I(P, Q) \geq 1$ with equality if and only if ρ is the density matrix of a coherent state $|z'\rangle\langle z'|$. Moreover, one has

$$\ln\left(\frac{e}{\hbar}\Delta_\mu q \Delta_\mu p\right) \geq I(Q) + I(P) \geq I(P, Q). \qquad (4.5)$$

The corresponding variances are

$$\left(\Delta_\mu q\right)^2 = \left(\Delta_\rho q\right)^2 + \sigma_q^2; \; \left(\Delta_\mu p\right)^2 = \left(\Delta_\rho p\right)^2 + \sigma_p^2 \qquad (4.6)$$

where Δ_ρ denotes the quantum mechanical variance and thus

$$\left(\left(\Delta_\rho q\right)^2 + \sigma_q^2\right)\left(\left(\Delta_\rho p\right)^2 + \sigma_p^2\right) \geq \hbar^2. \qquad (4.7)$$

Taking account of $\sigma_q \sigma_p = \frac{1}{2}\hbar$, by minimizing Eq. (4.7) over σ_q one gets the standard uncertainty relation $\Delta x \Delta p \geq \frac{\hbar}{2}$.

4.1.2 Kak's Approach to Quantum Information

In the case of the mixed state described by the matrix

$$\rho = \begin{bmatrix} p & 0 \\ 0 & 1-p \end{bmatrix}, \qquad (4.8)$$

the von Neumann entropy (4.1) becomes

$$S_n(\rho) = -p \ln p - (1 - p)\ln(1 - p) \qquad (4.9)$$

and the corresponding Shannon information is

$$I = p \ln p + (1 - p)\ln(1 - p) \qquad (4.10)$$

As a consequence, the von Neumann entropy of a pure state is zero; it means that the copies of the pure state does not carry any information. On the other hand, an unknown pure state can be used to transmit information, so we can consider von Neumann limited for the modern exigencies of Quantum Information technologies.

If one considers the information transfer problem from the point of view of the preparer of the state and the experimenter, it is clear that both mixed and pure states provide information to the experimenter. In this regard, a detailed analysis has been provided recently by Subhash Kak: *For a two-component elementary mixed state, the most information in each measurement is one bit, and each further measurement of identically prepared states will also be one bit. For an unknown pure state, the information in it represents the choice the source has made out of the infinity of choices related to the values of the probability amplitudes with respect to the basis components of the receiver's measurement apparatus. The maximum information in a pure state is thus infinite. On the other hand, each measurement of a two-component pure state can provide one bit of information. But if it is assumed that the source has made available an unlimited number of identically prepared states, the receiver can obtain additional information from each measurement until the probability amplitudes have been correctly estimated. Once that has occurred, unlike the case of a mixed state, no further information will be obtained from testing additional copies of this pure state. The receiver can do this by adjusting the basis vectors so that he gets closer to the unknown pure state. As the adjustment proceeds, the amount of information that he would obtain from each measurement will decrease. The information that can be obtained from such a state in repeated experiments is potentially infinite in the most general case. But if the observer is told what the pure state is, the information associated with the states vanishes, suggesting that a fundamental divide exists between objective and subjective information. [...] One can speak of information associated with a system only in relation to an experimental arrangement together with the protocol for measurement. The experimental arrangement is thus integral to the amount of information that can be obtained.* [2].

In order to conciliate the fact that, according to the von Neumann results, the entropy for an unknown pure state is zero with the fact that repeated measurements on copies of such a pure state do communicate information, Kak proposed a measure for the informational entropy of a quantum state that takes account of information in the pure states and the thermodynamic entropy. In Kak's approach, the quantum information emerges from an interplay between unitary and non-unitary evolution. In particular, the starting idea of Kak's model is that the informational entropy of a quantum system with density matrix ρ is given by relation

$$S_i(\rho) = -\sum_i \rho_{ii} \ln \rho_{ii}. \tag{4.11}$$

The entropy (4.11) indicates the average uncertainty that the receiver has in relation to the quantum state for each measurement. The value of $S_i(\rho)$ is the amount of entropy of the quantum system that is accessible to the receiver.

The informational entropy $S_i(\rho)$ has the following properties:

(1) $S_i(\rho) \geq S_n(\rho)$, and the two are equal only when the density matrix has only diagonal terms.
(2) $S_n(\rho)$ is obtained by minimizing $S_i(\rho)$ with respect to all possible unitary transformations.
(3) The maximum value of $S_i(\rho)$ is infinity, true for the case where the number of components is infinite.

Kak's entropy defined by Eq. (4.11) and satisfying the properties (1), (2) and (3) allows us to face in a coherent way also the pure states (contrary to the von Neumann's entropy) by taking into significant consideration each single state. It is also interesting to remark that the informational entropy introduced by Kak can resolve the puzzle of entropy increase in the Universe. It assumes that the Universe had immensely large informational entropy namely was in a low-entropy quantum state in the beginning, and that then, during the stages of the physical evolution, this informational entropy was transformed into thermodynamic entropy producing high-entropy quantum states, in part because of the second law of thermodynamics, in part because of the expansion of the universe, and above all because non-unitary evolutions of pure states. In this picture, one has both unitary, U, and non-unitary, M_i, or measurement, operators, and thus the density operator for each elementary state is:

$$|\phi\rangle_{new} = \left\{ \begin{array}{c} U|\phi\rangle \\ \dfrac{M_i|\phi\rangle}{\sqrt{\langle\phi|M_i^+ M_i|\phi\rangle}} \end{array} \right. \qquad (4.12)$$

where the first regards unitary evolution and the second non-unitary evolution. When the evolution is described only by non-unitary operators, the elementary state will change from the pure state $|\varphi\rangle = \alpha|0\rangle + \beta|1\rangle$ to the mixed state given by the density matrix:

$$G = \begin{bmatrix} |\alpha|^2 & 0 \\ 0 & |\beta|^2 \end{bmatrix} . \qquad (4.13)$$

In this case, the von Neumann entropy is finite because the informational entropy associated with (4.13) has transformed completely from that of the pure state to that of the mixed state.

The underlying hypothesis is that the primeval universe contained an high enthalpic order, ready to burst off energy and to produce structures in a process of entropy increasing, where the non-unitary operators play a key role.

4.1.3 Sbitnev's Approach to Quantum Entropy

As shown in Sect. 1.2.5 in the chapters (Refs. [36, 37] in Chap. 1) Sbitnev proposed the following definition of the quantum entropy

$$S_Q = -\frac{1}{2}\ln\rho \tag{4.14}$$

where ρ is the density of the ensemble of particles associated with the wave function under consideration. According to Sbitnev, the space-temporal distribution of the ensemble of particles describing the physical system is assumed to generate a modification of the geometry of the configuration space described by a quantum entropy given by Eq. (4.14). The quantum entropy (4.14) was introduced by Sbitnev to indicate the degree of order and chaos of the vacuum supporting the density of the ensemble of particles associated with the wave function. As we have already seen in Sect. 1.2.5, Sbitnev showed that, by starting from the quantum entropy (4.14) a new picture of non-relativistic de Broglie-Bohm theory emerges in which the quantum potential for a one-body system can be expressed as

$$Q = -\frac{\hbar^2}{2m}(\nabla S_Q)^2 + \frac{\hbar^2}{2m}(\nabla^2 S_Q) \tag{4.15}$$

The definition (4.15) implies that the quantum potential is a sum of two quantum corrector terms: the term $-\frac{\hbar^2}{2m}(\nabla S_Q)^2$ which can be interpreted as the quantum corrector of the kinetic energy $\frac{|\nabla S|^2}{2m}$ of the particle and the term $\frac{\hbar^2}{2m}(\nabla^2 S_Q)$ which can be interpreted as the quantum corrector of the potential energy. In fact, by introducing the quantum entropy (4.14), the energy conservation law in quantum mechanics for a one-body system reads

$$\frac{|\nabla S|^2}{2m} - \frac{\hbar^2}{2m}(\nabla S_Q)^2 + V + \frac{\hbar^2}{2m}(\nabla^2 S_Q) = -\frac{\partial S}{\partial t}. \tag{4.16}$$

In this picture, it is just the quantum entropy (4.14) the fundamental element determining that the quantum potential acts as an active information channel into the behaviour of the particles. The quantum entropy (4.14) describes the degree of order and chaos of the vacuum—a storage of virtual trajectories supplying optimal ones for particle movement—supports the density matrix ρ. Moreover, another important result of Sbitnev's quantum entropy lies in the fact that the path's distribution can be described by the matrix density and, in this regard, the two quantum correction terms of kinetic energy and potential energy both depending on the quantum entropy (4.14) modify the classical Feynman's path integral leading to a quantum complexified state space. When the quantum entropy (4.14) is negligible, or more in general

$$(\nabla S_Q)^2 \rightarrow (\nabla^2 S_Q) \tag{4.17}$$

(which can be considered as a correspondence principle for one-body systems inside Sbitnev's approach), the quantum dynamics will approach the classical dynamics and thus one obtains a classical information.

4.1.4 Quantum Entropy, Fisher Information and the Quantum Potential

As underlined by Carroll in (Ref. [1] in Chap. 3), the Fisher information can be considered another important element in order to study the relations between quantum potential, system's geometry and quantum information. The Fisher information can be interpreted as the information an observable random variable X carries about a not-observable parameter θ which the probability distribution X depends on. The Fisher information has arisen interest in relation to the study of the distributions of the observables of a quantum system.

As regards the physical meaning of the Fisher information and the attempts to connect Fisher information with the specific structural aspects of quantum mechanics the reader can find details, for example, in the references [2–6]. Here, we want to emphasize that Fisher information can be used to construct a metric able to connect the system's statistical outcomes and its global geometry and, in particular, to obtain the geometrodynamic, active information of the quantum potential.

In the recent paper *Weyl geometries, Fisher information and Quantum Entropy in Quantum Mechanics* [7] (as well as in the other paper [8]), the authors of this book showed that the quantum potential can be interpreted as a geometrodynamic entity indicating the deformation of the geometry and these features emerge by using Fisher information under a minimum condition in a picture based on a superposition of Boltzmann entropies. In this approach, the difference between classical and quantum information is similar to the difference between Euclidean and non-Euclidean geometry in the parameter space determined by the quantum entropy. In other words, the non-Euclidean geometry is here seen as the source of a quantum information that derives, on quantum level, from a quantum entropy which is defined as a vector of superposition of different Boltzmann entropies. By following the quoted references, here we want to outline the fundamental features of this approach.

By starting from an extension of the tensor calculus to operators represented by non-quadratic matrices, developed by G. Resconi (see, for a more detailed references:[8]), the authors introduced a new notion of geometric quantum information in a quantum entropy space in which the quantum potential expresses how the quantum effects deform the configuration space of processes in relation to the number of the systems microstates. In the case of a one-body system, the quantum potential is given by equation

$$Q = \frac{1}{2m} \left(\frac{1}{W_k^2} \frac{\partial W_k}{\partial x_i} \frac{\partial W_k}{\partial x_j} - \frac{2}{W_k} \frac{\partial^2 W_k}{\partial x_i \partial x_j} \right) \tag{4.18}$$

where the functions W_k define the number of microstates, which depend on the parameters θ of the distribution probability (and thus, for example, on the space-temporal distribution of an ensemble of particles, namely the density of particles in the element of volume d^3x around a point $\vec{x}$ at time t). The quantum potential (4.18) emerges as a consequence for the minimum condition of Fisher information

$$
\frac{\partial A}{\partial t} + \frac{1}{2m}p_i p_j + V + \frac{1}{2m}\left(\frac{1}{W_k^2}\frac{\partial W_k}{\partial x_i}\frac{\partial W_k}{\partial x_j} - \frac{2}{W_k}\frac{\partial^2 W_k}{\partial x_i \partial x_j}\right) =
$$
$$
\frac{\partial S_k}{\partial t} + \frac{1}{2m}p_i p_j + V + Q \qquad (4.19)
$$

obtained from the quantum action

$$
A = \int \rho \left[\frac{\partial A}{\partial t} + \frac{1}{2m}p_i p_j + V + \frac{1}{2m}\left(\frac{\partial \log W_k}{\partial x_i}\frac{\partial \log W_k}{\partial x_j}\right)\right] dt d^n x. \qquad (4.20)
$$

The quantum effects are here equivalent to a non-Euclidean deformation of the information space which is described by the following equation

$$
\frac{\partial}{\partial x^k} + \frac{\partial^2 S_j}{\partial x^k \partial x^p}\frac{\partial x^i}{\partial S_j} = \frac{\partial}{\partial x^k} + \frac{\partial \log W_j}{\partial x_h} = \frac{\partial}{\partial x^k} + B_{j,h} \qquad (4.21)
$$

where S_j is the quantum entropy given by the vector of the superposition of different Boltzmann entropies (one for each observer)

$$
\begin{cases}
S_1 = k \log W_1(\theta_1, \theta_2, \ldots, \theta_p) \\
S_2 = k \log W_2(\theta_1, \theta_2, \ldots, \theta_p) \\
\qquad \cdots \\
S_n = k \log W_n(\theta_1, \theta_2, \ldots, \theta_p)
\end{cases} \qquad (4.22)
$$

and

$$
B_{j,h} = \frac{\partial S_j}{\partial x_h} = \frac{\partial \log W_j}{\partial x_h} \qquad (4.23)
$$

is a Weyl-like gauge potential (see also [9]). In the geometric approach based on Eqs. (4.18–4.23), the difference between classical and quantum information corresponds to the difference between Euclidean and non-Euclidean geometry in the space parameters determined by the quantum entropy (4.22). Moreover, it is also interesting to introduce the quantity.[1]

$$
L_{\text{quantum}} = \frac{1}{\sqrt{\frac{1}{\hbar^2}\left(\frac{2}{W}\frac{\partial^2 W}{\partial x_i \partial x_j} - \frac{1}{W^2}\frac{\partial W}{\partial x_i}\frac{\partial W}{\partial x_j}\right)}} \qquad (4.24)
$$

which defines a typical quantum-entropic length that can be used to evaluate the strength of quantum effects and, therefore, the modification of the geometry with respect to the Euclidean geometry characteristic of classical physics. Once the

---

[1] In the next equations, for simplicity we are going to denote the generic function W_k with W.

quantum-entropic length (4.24) becomes non-negligible the system goes into a quantum regime.

It is also interesting to make a comparison of the geometric approach to quantum entropy and information analysed in this chapter to Sbitnev's as well as Kak's approach. If in Sbitnev's, by starting from the quantum entropy (4.14),the quantum potential emerges as an information channel into the behaviour of the particles given by relation (4.18) and thus associated with two quantum corrector terms of the energy of the system, in analogous way also in the approach here developed, the quantum potential emerges as an information channel associated with two quantum corrector terms of the energy of the system deriving from the functions W defining the number of microstates of the system (and thus also from the quantum entropy (4.22)). In fact, by equating Eqs. (4.18) and (4.15) one obtains

$$\frac{1}{2m}\left(\frac{1}{W^2}\frac{\partial W}{\partial x_i}\frac{\partial W}{\partial x_j} - \frac{2}{W}\frac{\partial^2 W}{\partial x_i\partial x_j}\right) = -\frac{\hbar^2}{2m}(\nabla S_Q)^2 + \frac{\hbar^2}{2m}(\nabla^2 S_Q). \qquad (4.25)$$

Equation (4.25) allows us to show that the two quantum corrector terms of the energy of the particle $-\frac{\hbar^2}{2m}(\nabla S_Q)^2$ and $\frac{\hbar^2}{2m}(\nabla^2 S_Q)$ of Sbitnev's quantum entropy can be related to two opportune quantum corrector terms of the approach here analysed depending on the functions W defining the number of the microstates of the system, namely $\frac{1}{2m}\left(\frac{1}{W^2}\frac{\partial W}{\partial x_i}\frac{\partial W}{\partial x_j}\right)$ and $\left(-\frac{1}{mW}\frac{\partial^2 W}{\partial x_i\partial x_j}\right)$ respectively:

$$\frac{1}{2m}\left(\frac{1}{W^2}\frac{\partial W}{\partial x_i}\frac{\partial W}{\partial x_j}\right) = -\frac{\hbar^2}{2m}(\nabla S_Q)^2 \qquad (4.26)$$

$$\left(-\frac{1}{mW}\frac{\partial^2 W}{\partial x_i\partial x_j}\right) = \frac{\hbar^2}{2m}(\nabla^2 S_Q) \qquad (4.27)$$

Thus, on the basis of Eqs. (4.26) and (4.27), one can say that the functions W appearing in the vector of the superposed Boltzmann entropies (4.22) are linked with the quantity (4.14) and thus with the density of the particles associated with the wave function. Moreover, by introducing the two quantum correctors of the energy given by Eqs. (4.26) and (4.27), the quantum Hamilton–Jacobi equation of non-relativistic Bohmian mechanics reads as

$$\frac{|\nabla S|^2}{2m} + V + \frac{1}{2m}\left(\frac{1}{W^2}\frac{\partial W}{\partial x_i}\frac{\partial W}{\partial x_j} - \frac{2}{W}\frac{\partial^2 W}{\partial x_i\partial x_j}\right) = -\frac{\partial S}{\partial t}. \qquad (4.28)$$

Equation (4.28) provides an energy conservation law in non-relativistic quantum mechanics, where the two quantum corrector (4.26) and (4.27) can be interpreted as a sort of modification of the geometry of the parameter space. On the basis of Eqs. (4.26), (4.27) and (4.28), one can say that the distribution probability of the wave function determines the number of microstates W, and a quantum entropy emerges from these functions W given by Eq. (4.22), and these functions

W, and therefore the quantum entropy itself (namely the vector of the superposed entropies), can be considered as the fundamental physical entities that determine the action of the quantum potential (in the extreme condition of the Fisher information) on the basis of Eq. (4.18).

In this way, inside the picture developed in this chapter, under the constraint of the minimum condition of Fisher information, the Boltzmann entropies defined by Eq. (4.22) emerge as "informational lines" of the quantum potential. In other words, each of the entropies appearing in the superposition vector (4.22) can be considered as a specific information channel of the quantum potential.

However, it must be emphasized that there is a fundamental difference between the geometric approach to quantum entropy and information analysed in this chapter and Sbitnev's approach. In fact, while in Sbitnev's approach the quantum entropy and information associated with the quantum potential emerge directly from the density matrix, instead here the quantum potential emerges from a minimum condition of Fisher information in a picture where the conventional Schrödinger equation is not taken as a fundamental concept. Here, the quantum potential is obtained as a gauge condition on the Boltzmann entropies of different observers, namely all observes must agree that there is a non-local "difference" from the classical case. While in Sbitnev's approach the non-local nature of quantum information emerges directly from the matrix density, instead here it emerges in a geometric picture where one has a specific Boltzmann entropy for each observer and therefore the primary, fundamental source of quantum information is not the matrix density but the deformation of the geometrical properties of the parameter space determined by the vector of superposition of different Boltzmann entropies. The derivation of the two quantum corrector terms of the energy of the system is different in the two approaches. But, as regards the feature of the quantum potential as information channel associated with a quantum entropy, the results are similar.

In them both, there's a common and extremely interesting clue for the Quantum Information. Kak underlines that we can speak about information only in relation to a specific experimental apparatus. On the other hand, it is just the last one to fix the form of the parameter space. All that suggests a new kind of morphogenetic quantum information, based on the Quantum Potential "deformations" and the qbits collective behaviour [10, 11].

4.2 Chaos and the Quantum Potential

Quantum chaos can be defined as the dynamical behaviour of the quantized version of a given classical Hamiltonian having a chaotic nature. It is quantitatively measured by the Lyapunov spectrum of characteristic exponents which represent the principal rates of orbit divergence in phase space, or alternatively by the Kolmogorov-Sinai invariant, which quantifies the rate of information production by the dynamical system.

The idea of using Bohm trajectories to study chaos was first proposed by Dürr et al. [12], and was mentioned in the books *The undivided universe* by Bohm and Hiley (Ref. [10] in Chap. 1) and *The quantum theory of motion* by Holland (Ref. [8] in Chap. 1). During the 1990s this subject was analysed and discussed by several authors [13–25]. In particular, in the classic paper *Quantum chaos in terms of Bohm trajectories* [24], Hua Wu and D. W. L. Sprung showed that, inside Bohm's formulation of quantum mechanics, a one-dimensional system is not chaotic, while a generic two-dimensional system may be chaotic, proposing that the quantum vortex is the main factor driving chaotic motion. By following the treatment in [23], let us consider a one-dimensional bound problem in which a and b are two points on the x-axis, with $a < b$, which move according to equation

$$\frac{d\vec{r}}{dt} = \frac{\nabla S}{m} = \frac{\hbar}{m}\text{Im}(\nabla \ln \psi) \tag{4.29}$$

arriving at $a(t)$ and $b(t)$. By using the continuity equation

$$-\frac{\partial R^2}{\partial t} = \nabla \cdot \left(R^2 \frac{\nabla S}{m} \right), \tag{4.30}$$

the probability $P(a, b) = \int_{a(t)}^{b(t)} R^2 dx$ on the interval between the two moving points is conserved:

$$\frac{dP(a,b)}{dt} = R^2(b(t),\, t)v_b - R^2(a(t),\, t)v_a + \int_{a(t)}^{b(t)} \frac{\partial R^2}{\partial t} dx = 0. \tag{4.31}$$

On the basis of Eq. (4.31), taking into account that $R^2 \geq 0$, $a(t) < b(t)$ and thus that orbits never cross, one obtains that a Bohm orbit is periodic with period T for a one-dimensional bounded problem if R^2 is periodic with period T and has only isolated zeros (which is the condition for periodic Bohm orbits). Now, if one considers two points that at time t are separated by an infinitesimal distance $\delta a(t)$, since in this case the conservation of probability can be written as

$$\delta a(0)R^2(a(0),\, 0) = \delta a(t)R^2(a(t),\, t), \tag{4.32}$$

the Lyapunov exponent for an orbit commencing at $a(0)$ is

$$\lambda(a(0)) = -\lim_{T \to \infty} \frac{1}{T}\ln\frac{R^2(a(T),\, T)}{R^2(a(0),\, 0)} \tag{4.33}$$

By considering an initial state where all the probability is concentrated in a narrow region (being a delta function the limiting situation), since by following a Bohm orbit the wave function amplitude $R(x(t),\, t)$ does not take infinity as its long term limit and since for large times T one has

$$R^2(x(T),\, T) < R^2(x(0),\, 0)\exp(-\varepsilon T) \tag{4.34}$$

for $a < x(0) < b$ and, from the conservation of probability

$$[b(T) - a(T)] \geq \frac{\int_a^b R^2(x, 0)dx}{\max[R^2(0)]} \exp(\varepsilon T) \qquad (4.35)$$

would grow without limit, one obtains that the quantity (4.33) is zero for a bounded system. This means that for a one-dimensional problem confined to a finite interval, the Bohm orbital motion is not chaotic.

As regards two-dimensional motion, in the paper [24] Hua Wu and D. W. L. Sprung studied the case of the rectangular billiard for which the energy levels are

$$E_{n_x, n_y} = \frac{\hbar^2 \pi^2}{2m} \left(\frac{n_x^2}{a^2} + \frac{n_y^2}{b^2} \right) \qquad (4.36)$$

namely

$$E_{n_x, n_y} = \frac{\hbar^2 \pi^2}{2ma^2 P^2} \left(P^2 n_x^2 + Q^2 n_y^2 \right) \qquad (4.37)$$

where a and b are the lengths of the billiard's sides, $b/a = P/Q$, with P prime to Q. When b/a is rational, the wave function is periodic with period not exceeding $T = \frac{8ma^2 P^2}{h}$. When b/a is irrational, the energy levels are incommensurate, both P and Q tends to infinity, so the maximal period tends itself to infinity, the wave function is not periodic and neither are the Bohm orbits. As regards chaotic behaviour, one aspect is linked to the geometry associated with the ratio of the two sides of the rectangular billiard. A chaotic orbit for a rectangular billiard can be evidenced by considering the Lyapunov exponents

$$\sum_i \lambda_i = -\lim_{T \to \infty} \frac{1}{T} \ln \frac{R^2(\vec{r}(T), T)}{R^2(\vec{r}(0), 0)} \qquad (4.38)$$

which leads to the conclusion that the long time average of the rate of expansion of a small area element is zero as a consequence of the conservation of probability. For a two-dimensional problem, $\lambda_2 = -\lambda_1$ and, if one chooses for example the starting position of the orbit at $(0.5, 0.2)$, evaluating the largest Lyapunov exponent, one obtains that the rate at which a near orbit diverges corresponds on average to doubling an initial perturbation for every 3.6 times the length a the particle moves. Here, the chaotic behaviour is tightly linked to the quantum potential: the more the energy is increasing, the more the wave function is complicated and the more the orbits are convoluted. Moreover, Hua Wu and D. W. L. Sprung suggested that orbits near a quantum vortex (namely near an isolated wave function node and its vicinity) diver quickly determining thus the most important factor responsible for chaotic orbits.

The first quantitative study of chaos as phenomenon generated by the motion of quantum vortices was provided by Wisniacki and Pujals in the paper [25]. These authors showed that the movement of quantum vortices—which result from wave function interferences and have no classical counterpart—implies chaos in the dynamics of quantum trajectories as a consequence of the intricate form of the quantum potential for these problems.

More recently, interesting results about the relation of chaos to the effects of vortices in the context of the quantum potential approach have been obtained by Contopoulos, Delis and Efthymiopoulos (see for example, [26–28]). Their main result lies in the fact that quantum chaos is due to the formation (by moving nodal points) of "nodal point-X-point complexes". More specifically, they found that near every nodal point there is a hyperbolic point, called X-point, that has two unstable directions, opposite to each other, and two stable directions, opposite to each other. In this picture, chaos is introduced when a trajectory approaches such an X-point and turns out to be stronger when the X-point is closer to the nodal point. The existence of an X-point close to a nodal point derives from the topological properties of the quantum flow holding in the neighbourhood of nodal points for an arbitrary form of the wave function. Efthymiopoulos and his colleagues made a theoretical analysis of the dependence of the Lyapunov characteristic numbers of the quantum trajectories on the size and speed of the quantum vortices. Moreover, they showed that trajectories that never approach a "nodal point - X point complex" are ordered. In particular, the recent paper *Order in de Broglie-Bohm mechanics* [29] focuses the conditions for the existence of regular quantum trajectories, namely trajectories with a zero Lyapunov characteristic number, obtaining a partial order even in some cases of chaotic trajectories. They demonstrated that there are trajectories avoiding close encounters with the moving nodal points, in an example of a quantum state consisting of a superposition of three eigenfunctions in a two-dimensional harmonic oscillator. Due to this effect, there can emerge regular quantum trajectories which extend spatially in a domain overlapping the domain of nodal lines. Then, they studied the influence of order in the so-called quantum relaxation effect, namely the approach in time of a spatial distribution ρ of an ensemble of particles following quantum trajectories to Born's rule $\rho = |\psi|^2$, even if initially $\rho_0 \neq |\psi_0|^2$, showing that relaxation can be removed when one considers ensembles also of chaotic trajectories, as a consequence of the fact that the existence of order in a system obstructs the extent to which chaotic trajectories can mix in the configuration space.

Finally, as regards the link between the quantum potential and chaotic orbits, relevant perspectives are also suggested by the paper [30] by R. Parmenter and A. Di Rienzo. Considering a many-body system, a vanishing quantum potential in the quantum Hamilton–Jacobi equation is equivalent to adding the term

$$\frac{\hbar^2}{2m} \exp\left(\frac{iS}{\hbar}\right) \nabla^2 R = \frac{\hbar^2}{2m} |\psi|^{-1} \nabla^2 |\psi| = -Q\psi \tag{4.39}$$

to the Schrödinger equation so that the effective Hamiltonian becomes

$$H_{eff} = -\frac{\hbar^2}{2m} \nabla^2 + V + \frac{\hbar^2}{2m} |\psi|^{-1} \nabla^2 |\psi| \tag{4.40}$$

Since the quantity (4.40) depends on ψ the superposition principle is no longer valid. When $\phi \neq \psi$ one has

$$\int \left(\phi^* H_{eff} \psi - \psi H_{eff} \phi^* \right) d\tau = \frac{\hbar^2}{2m} \int \phi^* \psi \left[|\psi|^{-1} \nabla^2 |\psi| - |\phi|^{-1} \nabla^2 |\phi| \right] \neq 0$$

(4.41)

and thus H_{eff} is not Hermitian. Then one obtains

$$\frac{\partial}{\partial t} \int |\psi - \phi|^2 d\tau = \frac{\hbar^2}{2m} \int i[\psi^* \phi - \phi^* \psi] \left[|\psi|^{-1} \nabla^2 \psi - |\phi|^{-1} \nabla^2 \phi \right] d\tau \neq 0$$

(4.42)

On the basis of Eq. (4.42), if one takes under consideration the case where two initial conditions for the time dependent Schrödinger equation differ only infinitesimally, one obtains that, with the increasing of time, the two corresponding wave functions become quite different, indicating the possibility of deterministic chaos. Instead by removing the term $\frac{\hbar^2}{2m} |\psi|^{-1} \nabla^2 |\psi|$ from Eq. (4.42), the resulting Hamiltonian becomes Hermitian and the normalization of $(\psi - \phi)$ is time independent, so there can be no deterministic chaos. One can therefore say that, on the basis of Parmenter's and Di Rienzo's approach based on Eqs. (4.39–4.42), the quantum potential can be interpreted as a constraining force preventing deterministic chaos.

References

1. Lieb, E.: Proof of an entropy conjecture of Wehrl. Commun. Math. Phys. **62**, 35–41 (1978)
2. Kak, S.: Quantum information and entropy. Int. J. Theor. Phys. **46**(4), 860–876 (2007)
3. Frieden, B.R.: Science from Fisher information: a unification. Cambridge University Press, Cambridge (2004)
4. Hall, J.W.M.: Quantum properties of classical Fisher information. Phys. Rev. A **62**(1), 012107 (2000)
5. Shun-Long, L.: Fisher information of wavefunctions: classical and quantum. Chin. Phys. Lett **23**(12), 3127–3130 (2006)
6. Luati, A.: Maximum Fisher information in mixed state quantum systems. Ann. Stat. **32**(4), 1770–1779 (2004)
7. Fiscaletti, D., Licata, I.: Weyl geometry, Fisher information and quantum entropy in quantum mechanics. Inter. J. Theor. Phys. **51**(11), 3587–3595 (2012)
8. Resconi, G., Licata, I., Fiscaletti, D.: Unification of quantum and gravity by non classical information entropy space. Entropy **15**, 3602–3619 (2013)
9. Castro, C., Mahecha, J.: On nonlinear quantum mechanics, Brownian motion, Weyl geometry and Fisher information. Prog. Phys. **1**, 38–45 (2006)
10. Licata, I.: Effective physical processes and active information in quantum computing. Quant. Biosyst. **1**, 51–65 (2007) (Reprinted in New Trends in quantum information, Licata, I., Sakaji, A., Singh, J.P., Felloni, S. (eds.) Aracne Publ, Rome (2010))
11. Licata, I.: Beyond turing: hypercomputation and quantum morphogenesis. Asia Pac. Math. Newslett. **2**(3), 20–24 (2012)

12. Dürr, D., Goldstein, S., Zanghi, N.: Quantum chaos, classical randomness, and Bohmian mechanics. J. Stat. Phys. **68**, 259–270 (1992)
13. Schwengelbeck, U., Faisal, F.H.M.: Definition of Lyapunov exponents and KS entropy in quantum dynamics. Phys. Lett. A **199**, 281–286 (1995)
14. Parmenter, R.H., Valentine, R.W.: Deterministic chaos and the causal interpretation of quantum mechanics. Phys. Lett. A **201**, 1–8 (1995)
15. Faisal, F.H.M., Schwengelbeck, U.: Unified theory of Lyapunov exponents and a positive example of deterministic quantum chaos. Phys. Lett. A **207**, 31–36 (1995)
16. Iacomelli, G., Pettini, M.: Regular and chaotic quantum motion. Phys. Lett. A **212**, 29–38 (1996)
17. Sengupta, S., Chattaraj, P.K.: The quantum theory of motion and signature of chaos in the quantum behavior of a classically chaotic system. Phys. Lett. A **215**, 119–127 (1996)
18. Parmenter, R.H., Valentine, R.W.: Properties of the geometric phase of a de Broglie-Bohm causal quantum mechanical trajectory. Phys. Lett. A **219**, 7–14 (1996)
19. Dewdney, C., Malik, Z.: Measurement, decoherence and chaos in quantum pinball. Phys. Lett. A **220**, 183–188 (1996)
20. Oriols, X., Martìn, F., Suñé', J.: Implications of the noncrossing property of Bohm trajectories in one-dimensional tunneling configurations. Phys. Rev. A **54**, 2594–2604 (1996)
21. Parmenter, R.H., Valentine, R.W.: Chaotic causal trajectories associated with a single stationary state of a system of non-interacting particles. Phys. Lett. A **227**, 5–14 (1997)
22. De Alcantara Bonfim, O.F., Florencio, J., Sa Barreto, F.C.: An approach based on Bohm's view of quantum mechanics quantum chaos in a double square well. Phys. Rev. E **58**, 6851–6854 (1998)
23. Goldstein, S.: Absence of chaos in Bohmian dynamics. Phys. Rev. E **60**, 7578–7579 (1999)
24. Wu, H., Sprung, D.W.L.: Quantum chaos in terms of Bohm trajectories. Phys. Lett. A **261**, 150–157 (1999)
25. Wisniacki, D., Pujals, E.: Motion of vortices implies chaos in Bohmian mechanics. Europhys. Lett. **71**, 159 (2005)
26. Efthymiopoulos, C., Contopoulos, G.: Chaos in Bohmian quantum mechanics. J. Phys. A: Math. Gen. **39**, 1819–1852 (2006)
27. Efthymiopoulos, C., Kalapotharakos, C., Contopoulos, G.: Nodal points and the transition from ordered to chaotic Bohmian trajectories. J. Phys. A **40**, 12945–12972 (2007)
28. Efthymiopoulos, C., Kalapotharakos, C., Contopoulos, G.: Origin of chaos near critical points of quantum flow. Phys. Rev. E. Stat Nonlin. Soft. Matter Phys **79**, 1–18 (2009)
29. Contopoulos, G., Delis, N., Efthymiopoulos, C.: Order in de Broglie-Bohm mechanics. J. Phys. A.: Math. Theor. **45**, 5301 (2012)
30. Parmenter, R.H., DiRienzo, A.L.: Reappraisal of the causal interpretation of quantum mechanics and of the quantum potential concept arXiv:quant-ph/0305183 (2003)